MELTDOWN

Discover Earth's Irreplaceable Glaciers and Learn What You Can Do to Save Them

Anita Sanchez

ILLUSTRATED BY
Lily Padula

Workman Publishing • New York

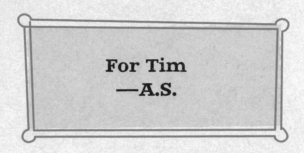

For Tim
—A.S.

Library of Congress Cataloging-in-Publication Data is available.

ISBN 978-1-5235-0950-8

Design by John Passineau
Photo research by Sophia Rieth

Photo credits
Alamy: Daniel J. Cox p. 50; NASA/Dembinsky Photo Associates p. 15 (top); Alexey Stiop p. 9 (top).
Anita Sanchez: p. 51. **Getty Images:** Zzvet/iStock 29 (top). **Jill Pelto:** p. 29 (bottom), p. 74.
Mauri S. Pelto: p. viii, p. 108. **Science Source:** Larry Landolfi p. 98; NASA p. 9 (bottom);
USGS/John Clemens p. 66. **Shutterstock:** Nicram Sabod p. 15 (bottom).

Workman books are available at special discounts when purchased in bulk for premiums and sales promotions as well as for fundraising or educational use. Special editions or book excerpts can also be created to specification. For details, contact the Special Sales Director at specialmarkets@workman.com.

Workman Publishing Co., Inc.
225 Varick Street
New York, NY 10014-4381

workman.com

Printed in China on responsibly sourced paper.
First printing November 2022

10 9 8 7 6 5 4 3 2 1

Contents

Foreword
by
Jill Pelto

What if you woke up in the morning on a mountain covered in ice and snow—a cool landscape of blue, white, and gray, even in late summertime? As you look down the mountain, you see warmer colors of red, orange, and purple as wildflowers bloom.

This is my experience working in the North Cascades, a mountain range in Washington state. For more than a decade, I've come here every year not just to look at the beautiful colors but to learn about the health of mountain glaciers—thick rivers of ice that slowly flow down slopes to create a landscape that looks like it was carved by a giant.

I grew up in the northeastern United States where there are no glaciers anymore, but I learned about them when I was very young. My dad, Dr. Mauri Pelto, whom you will learn more about later in this book, regularly travels to glaciers to measure how much our warming world is changing them. I remember being five years old and confused when he would travel across the country to do this work. He would return with many photos of himself on the snowy mountainside, but I always pictured him on a scary cliff of ice.

As I grew and spent more and more time outside, my curiosity about the natural world grew, too. My parents continued to teach my sister, brother, and me about Earth, and my dad told us that as soon as we were strong enough, we could come with him to study the glaciers in Washington. To do this, we had to be able to hike one hundred miles (160.93 km) up and down mountains and glaciers for three weeks.

In high school, after enough preparation—including a lot of exercise and a practice backpacking trip—I was ready to join my dad on his research trips to Washington. I remember my first trip to the glaciers really well because the landscape was entirely new to me. I didn't know there could be so many wildflowers, or that the deep snowbanks could last through the whole summer. It was amazing to see crevasses cut so deeply into the ice in beautiful patterns. It was hard work to hike and camp for such a long period of time, but the challenge really helped me grow and learn.

Since my first trip to Washington in 2008, I've gone every year to work on the North Cascade Glacier Climate Project. Founded by my dad in 1983, the project focuses on the same group of glaciers, where snowfall and glacial retreat are

Measuring Crevasse Depth, a watercolor by Jill Pelto, inspired by her work with the North Cascade Glacier Climate Project

measured each year. The purpose of the project is to have an annual record of change. Over the years, I've had the opportunity to visit glaciers in Antarctica, Canada, New Zealand, Scandinavia, and South America, as well.

My experience has taught me that glaciers are important for many reasons. They are freshwater reservoirs—in the warmer months, melting ice and snow flow into rivers and lakes and help keep them full. Especially in years of drought, the water from the glaciers is needed to sustain streams and reservoirs. Glaciers are also an important part of the ecosystem—for example, in Washington state, mountain goats use the ice and snow to travel quickly and keep cool. During warmer months, wildflowers soak up the melted ice and snow so they can bloom. All year round, people visit glaciers to find beauty and joy while hiking, climbing, and skiing.

By observing glaciers around the world, I've also witnessed climate change and glacial melt firsthand, from a large new pond formed from melting glacier ice to the polluted air caused by forest fires in the western United States, which impacts our research in Washington by making air quality and visibility very poor. As the world continues to warm, it's increasingly important to learn how our future is dependent on the health of glaciers and all the creatures who rely on them. We can learn about these topics by asking questions and looking for answers. That's what science is all about.

However, science is not the only way to learn about the world. Art is another path of exploration. Everywhere I travel, I bring art supplies and a sketchbook to draw or paint what I see. Whether I sit by a mountain lake to capture its depth of color, or quickly sketch a herd of mountain goats before they scamper away, my artwork inspires me to ask more questions about the environmental changes I see.

Science and art are two different ways to share what we know about the world, but I think it's important that they work together. Scientific research helps us measure what's happening, while art and storytelling help us communicate it to more people. In this book, you will explore glaciers through all of these ways: scientific data, art, and true stories of glacier expeditions. You will go on an adventure—like many of mine—and learn about what glaciers are, how they form, why they are changing, and why glacial melt matters to all of us, whether or not we live near a glacier.

In the last few years, so many young people like you have taught me what they know about climate change and why it matters. They've shared school projects about renewable energy and told me about their efforts to reduce plastic use at home. These stories give me hope for our future. We may be facing more environmental challenges than ever before—warming temperatures, melting glaciers, sea-level rise, and species extinction—but together we can raise our voices to take action and help save the planet we call home. I invite you to listen, learn, and act in whatever ways you can. This book will help you get started.

—JILL PELTO,
glaciologist and artist

Jill working on the Rainbow Glacier (Mt. Baker, WA) with the North Cascade Glacier Climate Project

About the Author

Anita Sanchez is the author of many children's science books that sing the praises of unloved plants and animals. For many years she worked as an environmental educator for the New York State Department of Environmental Conservation. She is now an author and freelance educational consultant to schools, libraries, and nature centers.

Introduction

Glaciers in Crisis

Imagine you're looking at our planet—a whirling blue-and-green marble—spinning in the blackness of outer space. Draw the focus closer and see the jagged mountain ranges, rivers, and deserts. Now look at the poles, where circles of ice cover both the north and south like frosty caps. Moving even closer, notice the long stripes of ice stretching between mountains and curving through valleys all over the world. These are glaciers. More than a hundred thousand of them cover 10 percent of the land on our planet.

Glaciers are places of stunning beauty that have formed over hundreds—sometimes even millions—of years. They flow over the land like massive frozen rivers with waves and waterfalls of ice. But they're far more than just scenic wonderlands—glaciers are part of ecosystems supporting a vast network of life. Just like oceans and forests, glaciers play a crucial role in the health of our planet.

These immense masses of ice cover large swathes of land and water like protective shields that reflect the sun's heat and cool the planet. They also hold almost three-quarters of Earth's fresh water and serve as a source of drinking water for millions of people. Outflow from melting glaciers is an essential source of water for many rivers, and it also cools oceans and changes ocean currents, which affect the world's weather. Glaciers provide an essential habitat for countless animals and plants that need cold in order to survive or that live in lakes and streams fed by meltwater. And glaciers play yet another unique role on Earth: Over a very long period of time, their movement (yes, glaciers move!) can literally create the mountains and valleys that form our planet's landscape. Whether you've ever seen a glacier or not, glaciers affect you: the land you walk on, the water you drink, and the weather around you.

But the glaciers are in trouble. Glaciers have always melted a bit in the summer as part of a natural annual cycle that has gone on for thousands of years. But now, because of climate change, glaciers are losing billions of tons of ice and snow every year, and the rate is accelerating fast. The frozen places of the world are vanishing before our eyes.

Already these changes are impacting every living thing on Earth— including you. Seas are rising. Weather patterns are growing more dangerous

and more intense. Critical water reserves are being lost. Many species are on the brink of extinction. In time, coastal cities will experience more and more flooding as climate refugees flee their homes for higher ground. And as glaciers get smaller, the effects will only get bigger.

Glaciers are a critical part of Earth's global ecosystem, so we have to save them in order to preserve the planet for future generations. And we have to act fast before time runs out.

How do we know the glaciers are melting? What is climate change and what's causing it? What does meltdown mean for the planet, and what will happen next if we don't change course? Most importantly, why should we care, and what can we do? This book offers answers to all of these questions.

First, let's explore the role that glaciers play in Earth's environment. Then we'll investigate how and why they're melting. We'll examine how the meltdown is already impacting life on Earth and what the future might look like. Along the way we'll learn from climatologists and glaciologists, as well as Indigenous peoples who have been keenly aware of the health of glaciers for centuries. Finally, we'll look at what we can all do to help save the glaciers and the planet— and how to get started.

Chapter 1

A World of Ice

The Role Glaciers Play on Our Planet

When you walk on a glacier, what do you see? Blue sky. Black mountains. A long slope of snow and ice glittering in the sun. The metal points of your crampons dig into hard-packed snow. *Crunch, crunch, crunch.* Each step is loud. When you stop walking, the only sound is the occasional croak of a raven, or the wind in the dark-green spruces. But listen closely: There's a *creeeak* and a *crack!* Under your feet, inch by inch, ice is creeping down from snowy mountains above you to the valley far below. It's like you're riding on the back of a big frozen giant. *Rrruummble. . .* A low grumble and mutter. Constant movement over the rough surface twists and bends the ice. As it strains under pressure, ice makes weird and terrifying noises. One second, you hear it whispering; then there's a *boom!* like a clap of thunder as a huge chunk breaks off.

You're walking on one of the most amazing places on Earth—a glacier.

A Long Time Ago . . .

Over eleven thousand years ago, our planet was a cold place.
Layers of ice, altogether more than a mile thick, stretched across the
earth from the North Pole halfway to the equator. Huge portions
of the planet—including much of what is now the United States—
were covered by glaciers. The places that are now New York City,
Chicago, and Seattle were all buried deep under tons of ice.
These glaciers would have been an awesome sight:
huge blue-and-white walls hundreds of feet high,
stretching to the horizon. They crept across the
land in frozen silence, broken by ear-
splitting crashes as avalanches
cascaded down.

Saber-toothed tigers and woolly mammoths roamed the frigid landscape, along with giant ground sloths and another, smaller mammal: humans.

This period of Earth's history was the most recent **ice age**, a long period when cooling temperatures result in the creation of glaciers that stretch across the planet. Since Earth was formed billions of years ago, there have been at least five major ice ages. Most scientists believe the ice ages were primarily caused by changes in Earth's orbit and the tilt of the planet's axis, changing the amount of sunlight the planet received.

While the most recent Ice Age ended thousands of years ago, our planet is still home to glaciers great and small.

What Are Glaciers?

A **glacier** is made of Earth's simplest ingredients: water, cold, and time.

When we think of snow, most of us picture it falling in winter and melting in spring. But where the climate is cold enough, like near the poles or in high mountain ranges, not all the snow disappears. And next winter, more falls on top.

As more and more snow piles up, the delicate points of the fallen snowflakes break off and the flakes turn into rounded, hard particles called **firn**. As years go by, heavy blankets of snow accumulate, squeezing out air from the layers below. Finally, the lowest layers of firn are compressed into solid ice. A glacier is born.

Glaciers have been a part of our world for countless millennia, and for as long as they have existed, they have been playing crucial roles on our planet—literally shaping, sustaining, and nourishing the earth.

> **{Glacier}** A large mass of ice that is formed over time from compacted layers of snow and that moves slowly over the land.

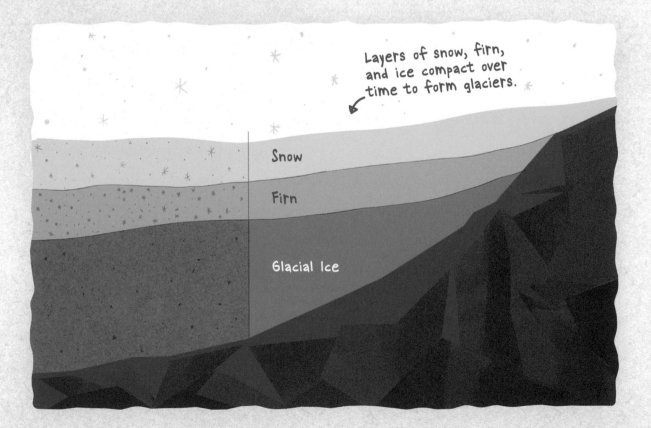

Layers of snow, firn, and ice compact over time to form glaciers.

Snow

Firn

Glacial Ice

RECORD-BREAKING GLACIERS

Want to climb one of the world's steepest glaciers? The Franz Josef Glacier in the Southern Alps of New Zealand stretches from a height of 9,800 feet (3,000 meters) down to almost sea level. You'd probably take a helicopter to get all the way to the top.

To explore the world's largest glacier, you'd have to travel to Antarctica. The Lambert Glacier, listed in the Guinness Book of World Records as the longest, is up to 60 miles (100 km) wide and 250 miles (400 km) long—more than twice the size of the state of New Jersey! It's hidden deep in the Antarctic, accessible only by special ships that can navigate through the iceberg-filled sea. Only a few humans have ever visited this remote place.

↖ Franz Josef Glacier

↙ Lambert Glacier

What Kind of Glacier?

There are two kind of glaciers—mountain glaciers and continental glaciers.

Mountain glaciers form anywhere the land is high enough and/or cold enough to have snow cover year-round. Some of them wind along gentle slopes and others cascade down steep-sided valleys. The high peaks of Nepal, China, and Russia are streaked with mountain glaciers. They're found in the jagged mountain range of the Andes in South America and flow down from the mountains of Iceland, Switzerland, Canada, and New Zealand. Glaciers even form near the equator—there's one on the top of Mount Kilimanjaro, Africa's highest mountain.

Continental glaciers are gigantic masses of ice that cover almost all of Greenland and Antarctica. While mountain glaciers are like rivers or waterfalls,

Mountain Glaciers ↘

Continental Glaciers ←

continental glaciers are more like giant pancakes oozing over thousands of square miles of land. When the ice reaches land's end, sometimes chunks called **icebergs** break off and crash into the sea.

A mountain glacier looks like part of the mountain it sits on, and you don't expect a mountain to move. But glaciers are always in motion, thanks to the unique and amazing properties of ice.

Seemingly hard as iron, ice can shatter like glass or stretch like rubber. It can even flow like water. If you hit an icicle with a hammer, it breaks. But if you attach a heavy weight to the end of the icicle, the ice will very slowly lengthen until the weight touches the floor. Ice has plasticity. That doesn't mean it's made from plastic—it means that the ice can change shape if there's enough pressure to make the bonds between the water molecules bend and stretch.

Glaciers Shape Our World

Since a glacier doesn't sit on a flat surface, gravity constantly pulls at it. Once the ice's weight is heavy enough, the glacier begins to flow along the ground beneath it, like a super-slow-motion river. Just as water always flows downhill, so do glaciers.

The movement of a glacier is, well, glacial—in other words, really, really slow. Too slow for impatient humans to notice. It's like watching a flower open, or the tide come creeping in. The Jakobshavn Isbræ glacier in Greenland is one of the fastest-moving in the world, but it's still not very fast, sometimes traveling as much as 150 feet (45.7 meters) in a day. Meanwhile, the continental glaciers of Antarctica might crawl along at the rate of a few inches per year.

But sometimes glaciers wake up and show their muscle. When a glacier flows over a steep ridge, the ice can break into blocks the size of houses, called **seracs**, that topple over with deafening crashes. And just as water ripples when it travels over rocks, the smooth surface of snow can rip, making cracks called

crevasses. Some crevasses are tiny, only inches deep. Others dive hundreds of feet deep into blue ice.

The most dangerous crevasses are hidden by a thin layer of snow. You can't see them—until you step on the fragile crust and it shatters beneath your feet!

WHY IS THE ICE BLUE?

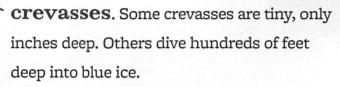

If you look into a crevasse, you'll see gorgeous shades of blue, ranging from sky blue to turquoise to sapphire. Not all ice looks blue, of course—most ice looks clear or white, because it has air bubbles in it. But when ice is under pressure, like the ice toward the bottom of the glaciers, most of the bubbles are squeezed out. This denser ice refracts light at the blue end of the spectrum. The oldest ice near the bottom of the glacier is the most compressed, so the bluer the ice, the older it is.

As glaciers move, they shape the land around them. Did you ever play in a sandbox and create mountains and valleys by pushing the sand around? Glaciers are like huge, icy fists shoving through the earth and reshaping it.

The weight of a glacier creeping inch by inch over the land can change the landscape as dramatically as a volcanic eruption or an earthquake—it just takes longer. During the ice ages, giant glaciers carved mountains and cliffs from solid rock, widened and deepened gorges, created lake basins, and etched fjords into coastlines. As the climate warmed and ice began to melt, giant **moraines** were left behind, creating hills and even islands, like Long Island in New York State. The Great Lakes, Norway's fjords, the Tien Shan range of China, and countless other parts of Earth's landscape were shaped by ancient ice. Even the hill your home sits on or the valley your school bus drives through might have been created by a glacier.

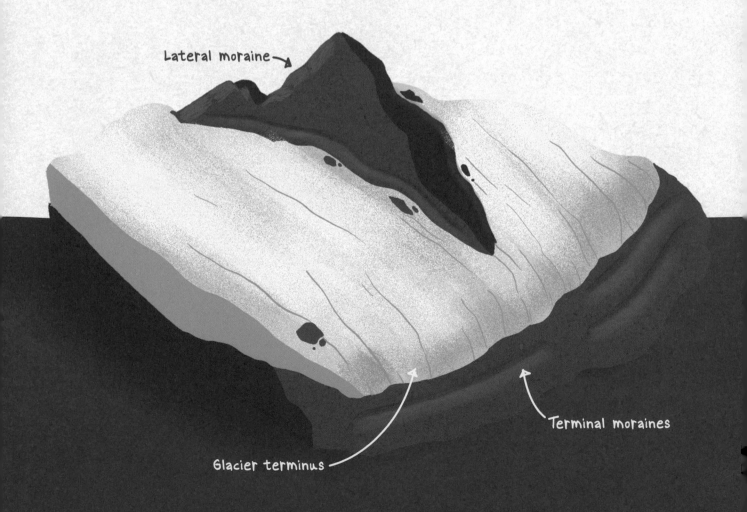

Lateral moraine →

Glacier terminus

Terminal moraines

The Great Lakes of North America were shaped by the movement of ancient glaciers.

Fjords in Norway →

MORAINES

As they move, glaciers act like bulldozers, scraping soil and rocks off the ground and pushing debris in front of them. Piles of gravel and rock moved by glaciers are called moraines. Piles of debris along the sides of the glacier are called **lateral moraines**. The ridge at the **terminus**, or bottommost end of the glacier, is called the **terminal moraine**.

Glaciers Are Life-Giving Reservoirs

For most of the year, glaciers hold billions of tons of water, frozen solid. But in the heat and dryness of late summer, glaciers seem to know it's time to water the land. During a melt, water gushes from the frozen reservoirs. Light reflecting off particles of silt in the water often gives it a beautiful blue-green color. This pure, clean water fills streams and rivers. It's icy cold and deeply refreshing to drink—a taste of melted snowflakes.

And there's a *lot* of drinkable water frozen into the world's glaciers. In fact, approximately two-thirds of all the fresh water on Earth is found in glaciers and ice caps. Glaciers are immense reservoirs of unpolluted water that humans desperately need.

People who live near glaciers in the Andes have relied on the local glaciers as a source of water since long before anyone can remember. The land is mostly desert, but for centuries local communities have been able to grow enough to feed themselves—as long as crops are watered by the annual cascades of glacial melt.

In the 1980s, farmers noticed that there was more and more water flowing from the glaciers. Few people saw the surge of extra water as a warning sign of increasing glacial melt—the extra water was hailed with enthusiasm. The Peruvian government used it to irrigate 100,000 more desert acres (40,500 hectares). Government-funded dams, canals, and hydroelectric plants were constructed. Thousands of people migrated to the area, building new towns and villages. Soon, so much produce was grown that it could be exported. Juicy blueberries and tender asparagus were shipped all over South America and flown to Europe and North America.

But as glaciers continue to shrink, the flow of water in the Andes is decreasing. If they continue to melt at their current rate, these glaciers will be gone by 2050. Soon the desert soil will be drier than ever.

And the people of the Andes are not the only ones depending on glaciers. In fact, one-sixth of the world's population lives in areas watered by glacier-fed rivers.

> **"Each year there is less water; each day there is less water."**
> —César Portocarrero, Peruvian climatologist studying glaciers in the Andes

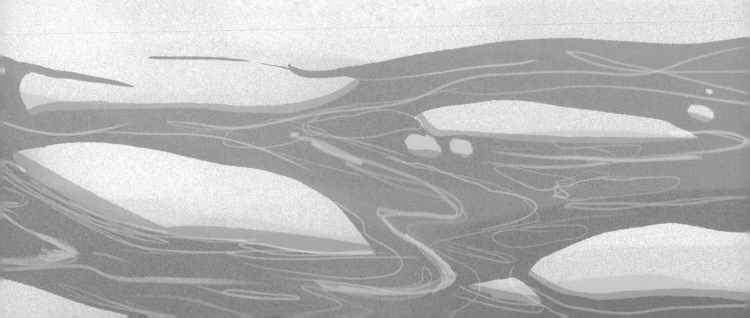

Glaciers Are Wildlife Habitats

Thousands of species of plants and animals are **chionophiles**, or "cold-loving organisms." Many kinds of wildlife, from mountain goats to ice worms, are found on glaciers. There's even a type of bird called the diuca finch, sometimes called the glacier bird, that can nest on the ice of high-altitude glaciers, incubating eggs in the bitter cold. Salmon and countless other species of fish depend on glacier-chilled water to survive, and these fish play key roles in food chains and ecosystems that support other species.

Diuca finch

All these animals, and many types of plants, depend on glaciers to provide the consistently cold habitat they need to survive. But what happens to these creatures as temperatures rise and the glaciers melt?

These hardy species aren't just cold lovers, they're cold *needers*. They've evolved to survive in the cold. Now they can't live without it. As rivers and oceans warm, whole fisheries are collapsing along with the human and wildlife communities that depend on them. All around the globe, countless species of wildlife are at risk of extinction as glaciers vanish.

Glaciers Are Homes

For many communities around the world, glaciers aren't only a source of life-giving water—they are homes, and they hold deep cultural significance.

Indigenous people from Canada to Peru and from Alaska to New Zealand have lived with glaciers for as long as tales reach into the distant past. Their daily lives were—and still are—deeply intertwined with the frozen landscape. For generations, they've depended on glaciers for water and as habitat for the animals they hunt.

But Indigenous people are also well aware that glaciers are dangerous places, constantly moving and changing, with sudden violent avalanches.

Among the Tlingit people of the Pacific Northwest, parents warned children never to speak disrespectfully of a glacier in case it might be listening. Glaciers were said to dislike noise, especially the voices of loud, boastful people. Even the aromas of cooking food could awaken a glacier's unpredictable wrath.

"In one place Alsek River runs under a glacier," a Tlingit Elder warned. "People can pass beneath in their canoes, but, if anyone speaks, while they are under it, the glacier comes down on them."

> "They say that in those times this glacier was like an animal, and could hear what was said to it."
> —DEIKINAAK'W, Tlingit Elder

The Inuit people of Canada have also lived with glaciers and ice for countless generations. In Canada's coldest region, Nunavut, it used to be that thermometers rarely rose above freezing, even in July. Sheila Watt-Cloutier, an Inuit climate activist, remembers growing up there, when dogsleds were the main way to travel across deep Arctic snow.

Now she can see her world melting away. Sled dogs stand paw-deep in puddles of muddy water. Fishermen die when they fall through once-thick sea ice that thins too early in the season. Soil that was once permanently frozen is crumbling as homes topple. Gushing water from melting glaciers causes dangerous floods.

Sheila is leading a movement of Indigenous people to demand action from the world's governments on the climate crisis. She's fighting fiercely to defend "the right to be cold."

Solid as Rock, Permanent as Mountains?

For uncountable millennia, glaciers have been a crucial part of the environment. Powered by tons of moving ice, these massive, frozen monsters are strong enough to reshape our planet. They seem solid as rock, permanent as mountains. How could they ever vanish?

You've heard the news—around the world, glaciers are melting. But isn't some melting normal, especially in summer when temperatures soar? Won't they regrow when winter comes around again?

It's true—ever since the last ice age, glaciers have had a summertime melt. Cold, fresh water flows across the parched soil, feeding dried-up streams and rivers. Then, in fall, as the cold weather creeps back, snow and ice in the uphill **accumulation zone** begin to replace what's been lost from the downhill **ablation zone**. In any given year the glacier might be a little smaller or larger

GLACIAL ZONES

The lower-altitude part of the glacier is called the **ablation zone**. This area naturally loses mass due to melting, evaporation, and chunks of ice breaking off. Higher up the slope is the **accumulation zone**, where new snow and ice pile up and then flow down to replenish what has been lost. The **equilibrium line** is where the amount of snowfall is equal to the amount of snowmelt in a year.

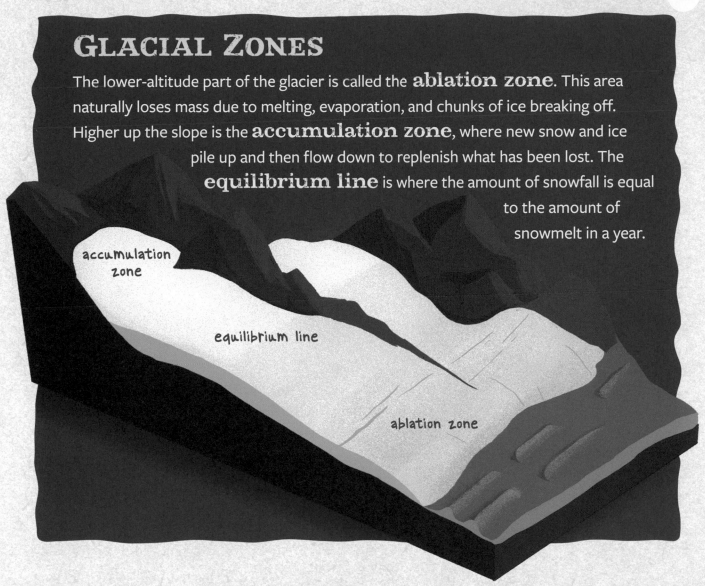

accumulation zone

equilibrium line

ablation zone

than the previous year, but over time its size remains more or less constant. This is all part of a natural annual cycle that has gone on for thousands of years.

But what we're seeing now isn't normal seasonal melting. In the last few decades—less than one human lifetime—an ominous change has occurred. Each year, summer is taking away more than winter can put back. The glaciers are disappearing—and fast.

How do we know? Where is the evidence for meltdown coming from? Let's explore the world of **glaciologists**—scientists who study glaciers—to observe firsthand what's happening to the frozen places of the world.

Chapter 2

Boots on the Ice and Eyes in the Sky

How We Know the Glaciers Are Melting

In the mid-1800s, a geologist named John Tyndall was awed by the glacier that covers the side of Mont Blanc, the tallest mountain in the Alps. It's called the Mer de Glace (the Sea of Ice) because it looks like a storm-tossed ocean, with giant waves frozen solid as they crest. Tyndall described it as "wild and wonderful turmoil," and he was fascinated by the glacier's frozen towers and pinnacles, its thunderous roars, and the deadly crevasses with blue light gleaming from their depths.

But he soon stopped sightseeing and got to work. "We did not come expressly to see it," he wrote. "We came to instruct ourselves about the glacier."

People who live near glaciers have always known that they move, but in the nineteenth century, some scientists refused to believe this was possible. During one of his visits to Mer de Glace, Tyndall devised an experiment to prove it. He put wooden stakes in a line across the width of the glacier, carefully measuring their exact location. After twenty-four hours, he observed with a thrill of delight that the stakes had moved several inches—and not all in a line. One was three times farther down the slope than others, revealing the hidden currents that move inside glaciers. He repeated the experiment the next day, and again a few days later, to confirm his results.

Tyndall's painstaking field measurements marked the beginning of careful, systematic recording and monitoring of glaciers. Many other scientists continued collecting data on glaciers, traveling to remote places of the earth and often encountering dangerous crevasses and risky avalanches. Eventually a worldwide group of scientists joined to form the International Glacier Commission in 1894.

This group developed into today's World Glacier Monitoring Service (WGMS). Through WGMS, which is sponsored by the United Nations, scientists all over the world have been studying and measuring glaciers and sharing their results with each other for over a century.

Global Annual Mass Change of Reference Glaciers

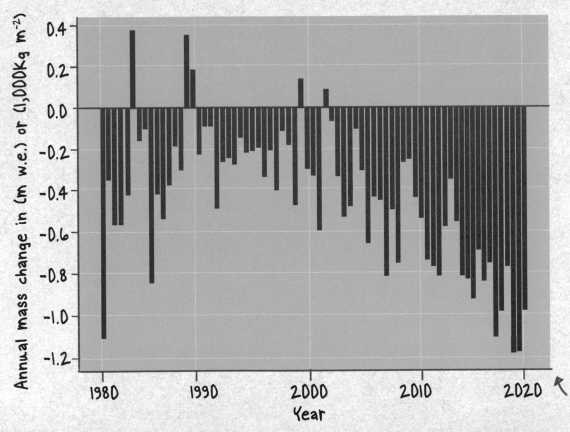

Benchmark glaciers have been steadily losing mass for decades.

{Glacial Mass Balance (GMB)} is a measurement that glaciologists use to evaluate changes in a glacier's size. GMB is determined by the total amount of snow and ice gained in the accumulation zone compared to the amount of snow and ice lost in the ablation zone. A glacier that has a negative balance will get smaller.

Dozens of glaciers worldwide are used as benchmarks so that changes can be closely tracked year after year. Glaciologists monitor these glaciers, measuring snow depth and snow melt, pinpointing terminus locations, and mapping changes in **Glacial Mass Balance**. They even still use Tyndall's technique of driving stakes into the ice to measure glacial movement. The scientists report their results to WGMS, which maintains a database of glacier measurements supplied by tens of thousands of researchers. This wealth of data on glacier size, movement, and change over time is freely available to anyone. Let's take a closer look at how glaciologists make these measurements.

FROZEN INTO THE ICE

In 1872, an Austrian explorer named Karl Weyprecht launched an expedition to the far north. Little was known about the polar region at the time—Weyprecht had no idea what he and his crew were sailing into.

As winter set in, ice began to form on the sea, trapping Weyprecht's ship. Stuck tight, there was nothing to do but wait for a thaw. After two years of waiting, they trekked across the frozen sea until they reached glacier-covered islands in the Russian Arctic and were finally rescued by Russian fishermen.

The experience left Weyprecht with a fascination for cold places. He suggested that nations all over the world work together to advance polar scientific research, and his idea caught fire. In 1882, scientists from many different countries joined to proclaim the first International Polar Year. An IPY is a year devoted to focused study of the polar regions. The fourth IPY was in 2007, and international scientists studied the Greenland and Antarctic ice sheets and Arctic mountain glaciers.

Boots on the Ice

Imagine you're a glaciologist hiking a glacier to assess how it might be changing from year to year. The glacier is a white and blue expanse of snow and ice, rimmed by jagged mountains. Only a few years ago the glacier filled the wide valley. But now it no longer reaches the valley's steep sides. The once-bulging surface of ice has flattened like a pricked balloon. A healthy glacier has a profile that bulges upward, or is convex, but the profile of this glacier is concave, with less mass in the middle than at the sides.

Convex

Concave

At the bottom of the slope, you observe that there are no crevasses. Since crevasses are caused by movement, this signals that the glacier is no longer moving forward. And the surface under your boots is mostly hard-packed ice, with little fresh snow covering it—another sign that something's wrong.

If you climb downhill toward the terminus where the glacier ends, you'll see that melted water from the glacier has made the ground wet and sodden, with sticky gray mud that clings to your boots. And listen: The sound of rushing water surrounds you. Trickles pour over mud and gravel, widening to a broad stream that gushes as loudly as a firehose. This isn't right. Some glacial melt is normal in the warmth of late summer, but this is far more than the usual amount of meltwater.

But the most alarming sign of all is the glacier's own footprint. As you plod down the muddy valley, you can see the ridges of moraines where the glacier has heaped up dirt and rocks. These moraines show you how far the glacier once traveled. But the outermost ridges are far away. You can see the terminal moraine that marks the place where the glacier once ended. It's hundreds of feet downslope from where the ice ends now, clearly showing how far the glacier has receded in recent years.

The relentless sun beats down on your back. There's a slow, ominous rumbling under your feet. Suddenly, part of the glacier's edge collapses, the snow and ice softening to a gush of white that flows like water down the mountainside. Ice vibrates under your feet as the avalanche plummets into the valley below with a mournful roar.

The glacier is talking—and it's getting louder. Its roars and groans tell of huge masses of ice breaking off, parts of its body that may never be replaced. This glacier is dying.

And it's not alone. All over the world, glaciers are shrinking away before our eyes. Many have already disappeared,

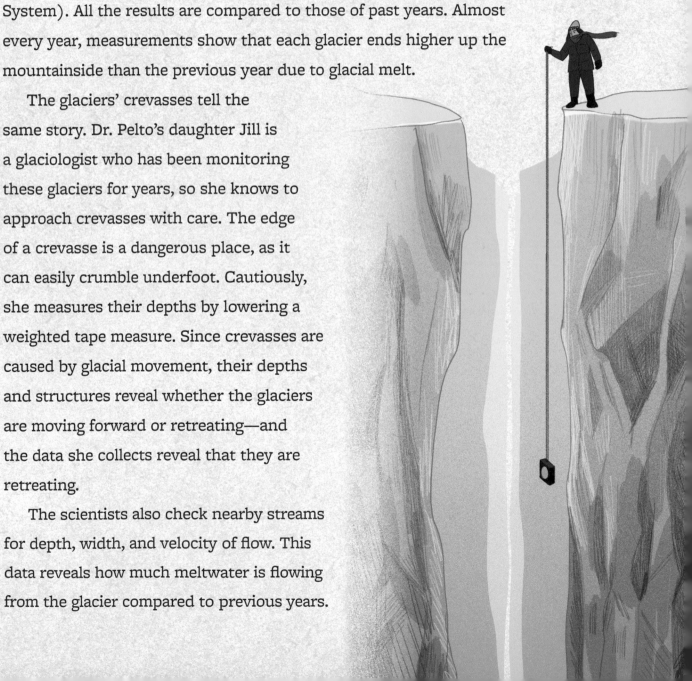

leaving behind only their footprint in gravel and mud—and glaciologists are witnessing these changes firsthand.

For four decades, Dr. Mauri Pelto has taken the same hike every August—up steep wilderness trails to the glaciers of North Cascades National Park in Washington State. Dr. Pelto's team of glaciologists visits the same glaciers annually, measuring them at the same time each year. They map each glacier's terminus, pinpointing the location and shape with a GPS (Global Positioning System). All the results are compared to those of past years. Almost every year, measurements show that each glacier ends higher up the mountainside than the previous year due to glacial melt.

The glaciers' crevasses tell the same story. Dr. Pelto's daughter Jill is a glaciologist who has been monitoring these glaciers for years, so she knows to approach crevasses with care. The edge of a crevasse is a dangerous place, as it can easily crumble underfoot. Cautiously, she measures their depths by lowering a weighted tape measure. Since crevasses are caused by glacial movement, their depths and structures reveal whether the glaciers are moving forward or retreating—and the data she collects reveal that they are retreating.

The scientists also check nearby streams for depth, width, and velocity of flow. This data reveals how much meltwater is flowing from the glacier compared to previous years.

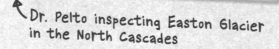

This glacial lake's blue-green color reveals the presence of glacial flour in the water.

If the water is cloudy, that means there are many fine particles of silt in the water. This silt, called **glacial flour**, is dust picked up by the glacier when it scraped the land long ago. When sunlight hits the silty water, it reflects a bright blue-green. This color is another vivid clue of glacial melt, showing that the parts of the glacier that once held this silt have melted.

The scientists can also literally feel the changes underfoot. While climbing on the glacier, Dr. Pelto wears metal **crampons** soled with two-inch-long (5.08 cm) spikes. Once he crunched through deep layers of snow. Now, more and more often as the years go by, his crampons dig into slopes of bare ice with no fresh snow. Warming temperatures mean rain, not snow, is falling on the glaciers. With no new snow in the accumulation zone, the glacier can't rebuild what it loses to melting.

Dr. Pelto inspecting Easton Glacier in the North Cascades

All of these signs and measurement point to the same alarming reality: Every single one of the forty-seven glaciers that Dr. Pelto's team has been monitoring is shrinking fast. In the almost forty years he's been watching, most have lost a third of their size. Four of the glaciers have completely disappeared.

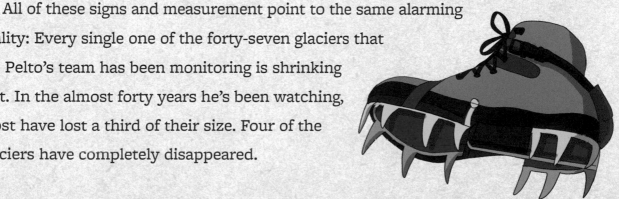

Eyes in the Sky

In addition to exploring glaciers up close, one of glaciologists' most effective tools for tracking changes is to watch glaciers from the sky.

Since the middle of the twentieth century, satellites have constantly circled our planet. Today's satellite-borne cameras take photographs of every inch of the earth. The Landsat program, managed by NASA and the US Geological Survey, is a series of earth-observing satellite missions that provide a constant stream of data about our planet. First launched in 1972, Landsat satellites capture images of every place on Earth every eight days. Other satellites carry a device called MODIS (moderate resolution imaging spectroradiometer). First launched in 1999, MODIS can see places every one to two days.

Satellite imaging lets scientists keep track of remote, hard-to-reach areas all across the planet. From the sky, they track forest fires, ocean currents, and cloud movement. They show where rainforests have been destroyed or new deserts have formed. And they help glaciologists track glaciers.

The World Glacier Monitoring Service has satellite images of many of the world's glaciers. Comparing photos of the same glacier taken a few years apart shows that glaciers are shrinking at a dramatic rate.

A Global Picture

Boots-on-the-ice research and images from satellites both tell the same story. Worldwide, glaciers are losing more than 300 billion tons (272 billion metric tons) of ice and snow every year.

And the rate of melt is speeding up. At the time of this writing, the world's glaciers are shrinking five times faster now than they were in the 1960s—far faster than scientists predicted. All over the world, from Greenland to Antarctica, the same thing is happening: meltdown. And once a glacier starts melting, it tends to keep on melting.

San Quintin Glacier Ice Loss

2001

2020

Between 2001 and 2020, San Quintin Glacier in Chile experienced significant ice loss (light blue), leading to the expansion of glacial lakes (dark blue).

11,000,000,000 TONS IN A DAY

During the summer of 2019, a heat wave broke records across Europe, leaving the glaciers bare, with no snow remaining from winter. The Greenland ice sheet lost eleven billion tons (nine billion metric tons) of ice—in a single day. Videos of foaming torrents pouring from Greenland's glaciers in the hottest July on record went viral. These powerful images made many people aware of how serious the climate crisis is.

Melting Leads to More Melting

Although it may have dark patches of gravel or dirt, a healthy glacier's surface is mostly light-colored, while the rocks and soil underneath are darker. Whenever a section of a glacier melts, the ground underlying it is uncovered. This dark ground absorbs the sun's heat, which raises the temperature of the air around it. So the glacier melts a little more. This melt reveals more dark rocks, which heat up, and so on. The more rock is exposed, the more the melt accelerates.

The same thing happens with glaciers near the sea. As pale sea ice disappears, darker sea water is exposed. The water absorbs heat, speeding up the melt of nearby ice even more.

A similar process plays out with the pinkish-looking patches on a glacier's surface called **watermelon snow**. The color is caused by algae that are cold-hardy enough to grow on ice crystals. The algae grow faster when temperatures are warmer. And since red is a darker color than white, watermelon snow melts faster than regular snow, exposing dark rocks. As the temperature rises, the algae spread faster. Once again, it becomes a vicious cycle of melting.

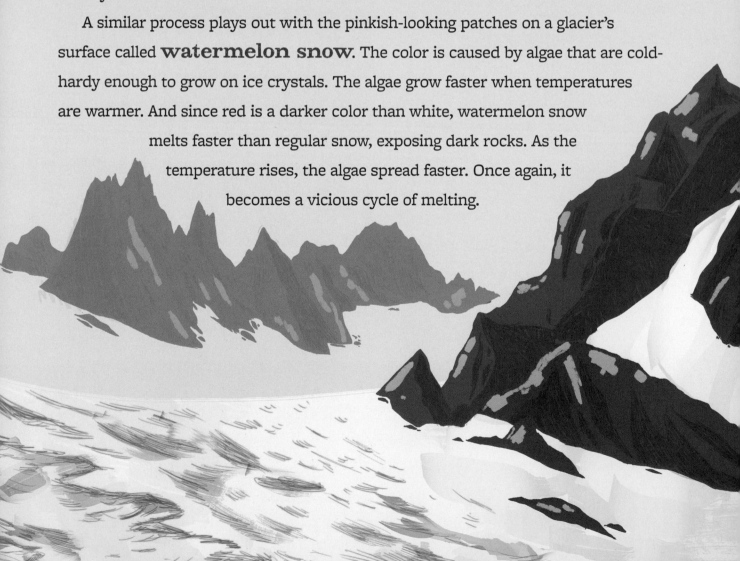

But here's the big question: Why is all this melting happening?

Ice is a glacier's strength. The unique properties of ice give the glacier the power to carve rock as well as the plasticity to flow like water. But ice is also its weakness.

Ice is a substance that exists very close to its melting point. This makes it different from other solids. For example, it takes a temperature of 2800°F (1510°C) to melt iron. Glass doesn't melt till it reaches a temperature of 2600°F (1427°C). But a temperature of just 33°F (0.56°C) will melt ice.

Glaciers depend on a very precise temperature to survive. The climate must be consistently below 32°F (0°C), the freezing point of water. In other environments, exact temperature doesn't make such an immediate difference. If the temperature in a desert, a prairie, or a forest warms by two degrees, not much changes, at least not right away. But if the temperature in glacial areas goes up, everything changes. Even a tiny rise above 32 degrees triggers melting.

And for many years now, temperatures all around the world *have* been rising. Something is causing our planet's climate to heat up. The very air we breathe is slowly, relentlessly changing.

> "It's a one-way road for these glaciers. Once they melt, they're not coming back in a foreseeable lifetime."
> —MAURI PELTO, US glaciologist

DARK COLORS ABSORB MORE LIGHT

On a sunny day, put your hand on a dark rock and then on a light-colored rock, and notice how much warmer the dark rock feels. Any dark object absorbs sunlight, changing it into heat. White objects reflect light, so they stay cooler.

Chapter 3

A Change in the Air

Why the Glaciers Are Melting

Weather reports used to be a mostly unexciting part of the daily news—rain showers, sunshine, maybe some snow. Nowadays, however, weather is making some of the world's most dramatic headlines: *Bomb Cyclone Hits California! Biggest Hurricane Ever Pounds Bahamas! Paris Swelters at 109 Degrees! Snow in Hawaii! Birds in Australia Drop Dead from Record Heat!*

More and more, these news stories contain words like *"Record-breaking!"* *"First time ever!"* and *"Historic!"* Like the melting glaciers, deadly storms and wild changes in weather patterns are symptoms of a change—a crisis—happening to our planet.

Climate change. We hear this scary term all the time. News stories feature it, politicians debate it, people all over the world are arguing over it. But what *is* climate change?

Weather vs. Climate

Look up at the sky. Is it sunny, partly cloudy, overcast? Is the sun shining or are snowflakes falling? That's the weather—the state of the atmosphere, the temperature, the rainfall at any given moment. Weather changes day to day—in fact, sometimes minute by minute.

Climate, on the other hand, is the overall weather pattern of an area, measured over a long time—years or even centuries.

A change in the weather isn't the same as a change in the climate. If it's cooler today than it was yesterday, that's a change in the weather. It doesn't mean the planet's climate is getting cooler.

To know that the climate is changing, we have to look at weather patterns across many years. Scientists have done just that. More than a hundred years of global thermometer measurements show that our climate is steadily getting warmer despite day-to-day fluctuations in weather.

But why?

> **{Climate change}**
> A significant change in patterns of weather over a long period of time.

A Nice, Warm Blanket of Air

Take a deep breath! Every human needs air to breathe—we can only live for minutes without oxygen. The thin, life-giving layer of air that surrounds our planet, called the **atmosphere**, includes the oxygen we need to survive, as well as other gases such as nitrogen, argon, hydrogen, and **carbon dioxide**.

We also need the atmosphere for another reason: to keep from freezing to death! Atmospheric gases surround Earth like an insulating blanket, trapping the sun's heat and keeping it from escaping back into space. This process is called the **greenhouse effect**. At the same time, glaciers, sea ice, and other

THE GREENHOUSE EFFECT

Two hundred years ago, when French scientist Joseph Fourier was trying to explain how the atmosphere kept Earth warm, he compared the atmosphere to a glass lid on a box that keeps heat trapped inside. His idea became known as the greenhouse effect.

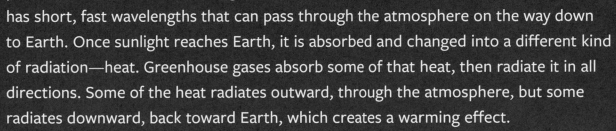

Like a glass-roofed greenhouse protecting plants, our atmosphere allows the sun's light to pass through and traps much of the sun's heat inside. Light and heat are examples of **radiation**, or energy that moves from one place to another in waves. Sunlight has short, fast wavelengths that can pass through the atmosphere on the way down to Earth. Once sunlight reaches Earth, it is absorbed and changed into a different kind of radiation—heat. Greenhouse gases absorb some of that heat, then radiate it in all directions. Some of the heat radiates outward, through the atmosphere, but some radiates downward, back toward Earth, which creates a warming effect.

snow-covered areas act as a crucial thermostat, helping to cool temperatures by reflecting heat from the sun instead of absorbing it as darker-colored rock or water would do.

Because of this, our planet's temperature remains steady, going up or down only slightly, within a range of hot and cold that plants and animals—including us humans—can survive.

Gases in the atmosphere that trap heat and hold it near Earth are called **greenhouse gases**. Carbon dioxide is especially powerful at influencing climate change because there's so much of it in the atmosphere.

Carbon Dioxide: Good or Bad?

You can't smell it. You can't see it or taste it. But we breathe carbon dixodide (CO_2) every minute of our lives.

Carbon dioxide has been floating around Earth's atmosphere for billions of years. All living things, including plants, animals, and other organisms like fungi need it to survive. Green plants absorb CO_2 molecules from the air to use in **photosynthesis**, their process of making food. Animals—including humans—rely on plants for our food, and we need to breathe in some CO_2. Our bodies can't function properly without it.

So some CO_2 in the air is a good thing. The problem starts when there's too much of it. Our atmosphere protects Earth like a blanket, but adding more CO_2 to the air is like adding too many extra layers of wool to the blanket. Whatever's inside gets hotter, and hotter. . . and the glaciers begin to melt. And that's exactly what's been happening: Vastly increased levels of CO_2 in the air are directly leading to climate change.

THE WOMAN WHO DISCOVERED CLIMATE CHANGE

Eunice Newton Foote was a scientist at a time when women weren't expected to be scientists. An American activist for women's rights in the mid-1800s, she was also an inventor and a scientist who conducted experiments at home. She is the first scientist known to have researched how sunlight affects atmospheric gases. In 1856, she proposed a theory that if carbon dioxide levels in the air rose, the planet would get warmer. Three years later, European physicist John Tyndall, a pioneer of glacier studies, announced similar conclusions. He did not credit Foote's initial discovery, though most likely he had never heard of her work, which wasn't widely published.

IT'S NOT ONLY CO₂

CO_2 isn't the only greenhouse gas that traps heat. Plain old water (H_2O) floating as a gas (called **water vapor**) absorbs the sun's heat and warms the earth. And the cycle keeps repeating.

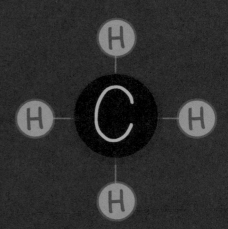

CO₂ molecular structure

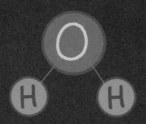

Warmer air holds more water vapor than cold air does. So as the temperature rises, air holds more water vapor. This traps more heat, raising the temperature so it holds yet more water vapor, trapping more heat, and on and on.

Methane (CH_4) is a gas that is given off when anything organic—something that was once alive, like plants or animals—decomposes. More methane forms when decomposition happens in a place where there isn't much oxygen, like a landfill. Landfills and garbage dumps create enormous amounts of methane. So do—of all things—cows! There are more than a billion cows on the planet, mostly due to the rise of the fast-food industry and the production of cheap hamburgers. When cows' waste decomposes, it gives off lots of methane. Methane levels are now twice as high as they were two centuries ago.

Nitrous oxide (N_2O) is another greenhouse gas with increasing levels. Because of the way its atoms are bonded together, a single molecule of N_2O has the heat-trapping potential of three hundred molecules of carbon dioxide. Most N_2O comes from nitrogen-based fertilizers used on farms and lawns.

These are the main greenhouse gases, but other air pollutants contribute to the greenhouse effect, too, from the fluorocarbons used in air conditioners and refrigerators to the black carbon given off by the diesel engines of trucks.

The Rise of Fossil Fuels

How does CO_2 end up in the air? With every breath, we produce CO_2. Animals and even plants exhale this gas. But CO_2 is also created when we burn materials that contain carbon, like wood or coal. The burning process combines the carbon with oxygen in the air, forming molecules of CO_2.

Now, if you lived long ago—let's say before the year 1700—you might burn wood to warm up your house in winter, or maybe cook over a wood fire. That would give off some CO_2 into the air. But going about your business three hundred years ago, you wouldn't cause nearly as much CO_2 to enter the atmosphere as people do today.

In fact, for thousands of years, most human activities—driving a wagon, making a clay pot, spinning wool by hand—didn't create much air pollution. Levels of CO_2 and other greenhouse gases in the air rose and fell over time, but emissions from humans made up only a small percentage of the total. Natural events like forest fires or volcanic eruptions gave off far more.

But in the middle of the 1700s, things began to change.

Over the course of only a few years, inventors came up with a variety of machines to make work easier, like spinning machines and mechanical weavers. This intense period of new inventions, called the Industrial Revolution, began in England and quickly spread all over the world. Before, people usually worked in their homes or in small workshops, slowly crafting items one at a time. Now, factory machines could produce almost anything, from cloth to shoes to paper, in record time. Machines could do things faster than people could—and cheaper, too.

And to power the rising number of new factories, there was an enormous increase in the burning of what we call **fossil fuels**. Petroleum (oil), coal, and natural gas have lain buried deep in the earth for hundreds of millions of years, but they aren't really fossils. They're the remains of plants and animals that lived during the Carboniferous Period, long before the time of dinosaurs. Fossil fuels are immense sources of energy, but since the cells of these ancient organisms still contain carbon, burning them gives off carbon dioxide.

The end of the nineteenth century was a time of head-spinning change as new inventions appeared fast. The first telephone call was made in 1876, and the first electric bulb lit up the night in 1879. The first automobile roared down the street in 1885, and within a few years, gasoline-powered engines began to replace horses as the world's main way to get from place to place. The first images captured on film by a motion-picture camera were released in 1893. The first manned airplane took off in 1903. The first fully synthetic plastic was invented in 1907. Change happened so fast that children who were born in lantern-lit log cabins could live to turn on their televisions and watch the first humans walk on the moon.

But there was a downside to all this exciting new progress. Smoke billowed from smokestacks of thousands of coal-powered factories. Exhaust fumes spewed from tailpipes of millions of gas-powered automobiles. And the CO_2 levels in the air rose dramatically.

Getting Warmer

At the beginning of the twentieth century, Swedish scientist Svante Arrhenius published a study revealing that global temperatures were rising due to the burning of fossil fuels. He was delighted at the thought of warmer weather. "We may hope to enjoy ages with more equable and better climates," he wrote happily in a later book. Longer summers, better crops, warmer winters—global warming sounded nice. And anyway, it all seemed to be far in the future. Few imagined that significant changes would come before hundreds, perhaps even thousands, of years.

But as time went on, the world's population skyrocketed. In 1700, there were approximately 600 million people on Earth. By 1900, there were more than twice as many: 1.5 billion. Now in the twenty-first century, more than seven billion people live on the planet. More people means more cars, more demand for electricity, more plastic, more fossil fuel use . . . and more CO_2 in the air.

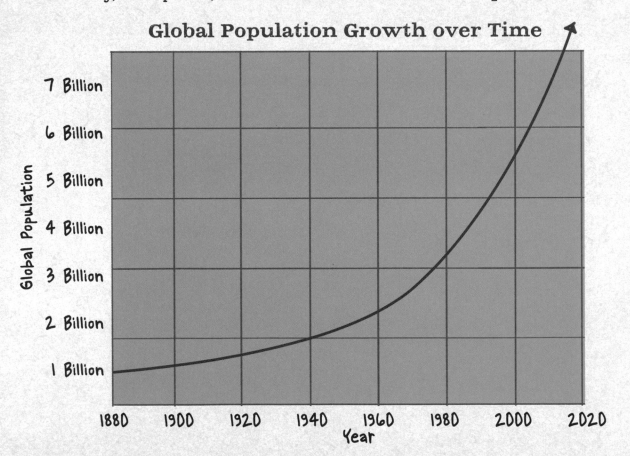

Global Population Growth over Time

Nature's Carbon Trap

As more and more humans have been putting more and more pollution into the air, we have also been destroying the perfect way to get rid of that pollution: trees. Like glaciers, forests are a critical part of Earth's regulatory system.

When we see a forest, we see a beautiful place of rustling green leaves and a rich wildlife habitat. What we don't notice is how trees are cleaning the air. Silently, invisibly, leaves are absorbing CO_2 through tiny pores on their undersides called stomata. During the food-making process of photosynthesis, carbon dioxide is broken down into the atoms that make it up: carbon and oxygen. The leaves release oxygen (which is handy for us humans, since we breathe it) and use carbon to produce sugars that plants use for food. Every single cell in a tree is partly made of carbon.

Plants are nature's carbon traps. Carbon is sequestered (removed from the atmosphere) when it becomes a part of a tree. And trees sequester a *lot* of carbon. More than a hundred tons of carbon can be absorbed and held by one acre of forest. The amount of carbon held in each tree varies—it depends on how old the tree is, how fast it grows, what species it is, and other factors. The oldest, biggest trees hold by far the most carbon. (Grasses and other green plants and algae trap carbon, too, though not as much because trees are so much larger.)

But for most of human history, civilizations have been destroying these natural carbon traps. Even back in the times of the Ancient Greeks, vast amounts of trees were cut for building material, for fuel, or to clear land for farming. As the centuries passed, this process—known as **deforestation**—has been repeated all over the world, from the vanished forests of ancient Greece to today's rainforests in Brazil.

And as the global population increased, deforestation increased, too. More people meant more demand for wood. And the technology used to cut down trees has gotten much more effective. Long ago, loggers used hand saws and axes. Today, chainsaws and motorized logging machines can clear an entire forest in hours. Billions of acres of forests have been destroyed, and the rate of loss is only getting faster.

Paper mills and lumberyards need more and more wood, and more land is constantly being cleared to build housing developments or shopping malls. Whole countries that were once densely wooded now have only tiny remnants of their original forests enclosed in parks. In the United States, before humans changed the landscape, a squirrel could travel tree to tree from Cape Cod to the Great Lakes and never set a paw to the ground;

east of the Mississippi, most of the continent was one big forest. Not anymore.

The huge rainforests of the Amazon in South America are called "the lungs of the earth" because they breathe out so much oxygen and sequester so much carbon. But these days, massive amounts of rainforests are burned or cleared to make way for farms and cattle ranches. Small family farms are often replaced by factorylike feedlots with thousands of cows confined in a small area. Many of these provide hamburgers for the fast-food industry.

The staggering scale of destruction is hard to imagine: worldwide, we lose many millions of acres of forest every year.

It's hard to believe a rainforest in the Amazon can affect a glacier thousands of miles away. But the loss of our forests has a devastating effect on our planet, especially glaciers and areas of sea ice. As forests are replaced by blacktops or houses or feedlots, Earth's natural carbon trap vanishes. Greenhouse gases in our atmosphere are increasing just as Earth's natural ability to absorb those gases is decreasing, raising Earth's temperature and leading to more glacial melt.

{Deforestation}
The complete removal of trees from a wooded area, turning it into open land. Deforestation causes soil erosion and habitat destruction.

DEFORESTATION IN ANCIENT GREECE

"What is left now is, so to say, like the skeleton of a body wasted by disease; the rich soft soil has been carried off and only the bare framework of the district left. . . . [Long ago] the plains of Phellus were covered with rich soil, and there was abundant timber on the mountains, of which traces may still be seen." The Greek philosopher Plato wrote these words mourning the disappearing forests of his homeland in about 360 BCE—more than two thousand years ago.

Bring on the Heat

For a long time, the slow warming of our climate went mostly unnoticed. When stories about climate change finally began to hit the national news in the 1980s, it was usually referred to as "global warming." And like the scientist Svante Arrhenius a hundred years earlier, people tended to think those words had a pleasant sound. In bitter New England sleet or a harsh Montana blizzard, who wouldn't enjoy a bit of global warming? We can keep the pool open longer, spring flowers will come earlier, and winter won't be so nasty!

Today, climatologists warn that this huge rise in greenhouse gases means that our atmosphere could warm up by three or four degrees in the coming decades. That might not seem like much of a change—what's a few degrees?

But as we've seen, just a few degrees of warmth has catastrophic consequences in the frozen parts of the world.

Reading the Future in the Past

We know that Earth's climate is warming and causing the glaciers to melt. But can we definitively say that rising greenhouse gas levels are to blame? After all, the earth has gone through periods of warming and cooling before—the ice ages, for instance. So how do we know that this current period of warming is different—that it's caused by human activity rather than natural fluctuations? In order to figure this out, we need to examine how CO_2 levels and temperatures have changed over time—very, very long periods of time.

Scientists are now keeping close track of greenhouse gas levels and temperature, but that's only been for a few decades. Even the oldest records only go back a few hundred years. From letters and diaries, we know that it was colder in past centuries—for example, in the Middle Ages people went ice-skating and

held festivals on the Thames River in London, which hasn't frozen solid since 1814. But we need to look farther back in time—*much* farther back.

Without inventing a time machine, how can anyone know exactly how much carbon dioxide was in the air ten thousand years ago, or how hot or cold it was? Fossils reveal many secrets of past eras, but they can't tell us about temperatures or CO_2 levels. What about trees? Each year, trees lay down a ring of cells in their trunks. You can age a tree by counting the rings and learn about past growing seasons, but even the oldest tree records go back only 12,000 years or so.

To study the climate of eons past, we need something that has persisted on Earth for a very long time. Something that can trap air samples and hold them for millennia. We need a time machine to bring us far, far back in history—a million years or more.

Fortunately, there are such time machines, deep in the coldest places of the earth. The secrets of past climates are hidden inside glaciers.

Chapter 4

Deep-Frozen
Time Machines

Glaciers' Hidden History

When you stand on a glacier, you're standing on top of history. Beneath your feet are layers that formed centuries ago, one year's snow piling up on the next.

If you were to get close to a crevasse and (cautiously) peer inside, you would glimpse stripes of lighter and darker ice on the walls of the crevasse. In summer, glacier snow softens and then refreezes. This makes large, coarse crystals that show up as lighter-colored layers. In winter, snow is densely packed with finer crystals, which form darker layers. You can even count these layers like tree rings. One dark band plus one light band equals one year. In this way, a glacier can be aged with accuracy.

Getting to the Bottom of History

To reach the oldest layers of a glacier, engineers have developed hollow drills that bite into hard ice and bore far down to extract long, thin samples called **ice cores**.

Before the first cores were drilled, no one realized how far back in time glaciers could take us. A small mountain glacier might contain ice that's merely hundreds of years old, but the continental ones are made of the thickest, oldest ice on

Earth. One of the first cores drilled in Greenland was four inches (10.16 cm) wide and a mile (1.6 km) deep, revealing ice from over 100,000 years ago! Several years later, coring began in Antarctica. Here they drilled "the great-grandaddy" of all ice cores—a cylinder of ice from a glacier that was *two miles* (3.22 km) deep. It contained ice from almost a million years ago.

And the time machine keeps bringing us farther and farther back. In 2017, glacial ice was discovered in another Antarctica core, estimated to be more than 2.5 million years old.

When you shine a light through the translucent cylinder of ice, you can see the summers and winters of history, piled up one on top of the other. In those layers are a wealth of information that help us visualize the climate of the past, understand the changes of the present, and make predictions about the future.

THE ICEMAN

Many of history's secrets are frozen into glaciers, and as the ice melts, some of these secrets are revealed. In 1991, hikers in the Ötztal Alps of Italy discovered a human body lying facedown in a melting glacier. They thought it was the victim of a recent accident, but when the body was examined, it turned out to be an unlucky wanderer who had died and been frozen into the ice—more than five thousand years ago.

The mummy was nicknamed "Ötzi." Researchers believe he was part of a group of nomadic mountain shepherds from the Copper Age, about 3300 BCE.

Ötzi was so well preserved by cold that his skin, teeth, and even his hair remained. He was a man in his forties who had been climbing the glacier in leather trousers and shoes made of thick bearskin. Stomach contents showed that his last meal was grain and antelope meat. He clutched a copper axe, and lodged deep in his left shoulder was a stone arrowhead. In a sheath, he carried a blood-stained knife.

Ötzi is now preserved in the South Tyrol Museum of Archaeology in Bolzano, Italy, in a display case kept at glacial temperatures. Thousands of visitors come every year to see the face of a man who lived five thousand years ago.

Blowing in the Wind

A glacier covered with new snow appears to be pure white, but mixed into the snow are dark specks blown onto the glacier by the wind. These dark specks are called **aerosols**.

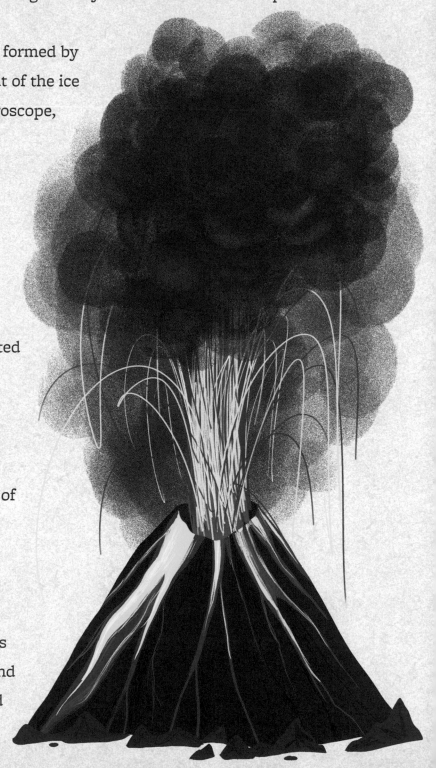

Every ice core has visible lines formed by aerosols. By cutting the specks out of the ice and examining them under a microscope, scientists can figure out what the specks are. They might be pollen from flowers that bloomed long ago, ash from a wood stove, soot from a coal-fired power plant or forest fire, or dust from an era of drought or erosion. When Mount Vesuvius erupted in 79 CE, it blasted tons of ash into the atmosphere. Winds carried the ashes all over the world, and some of them fell onto faraway glaciers. In ice cores from these glaciers, the black line of volcano ash shows clearly.

In many ice cores, thick, dark lines appear in layers of ice that formed after the start of the Industrial Revolution. These layers are heavily discolored with soot and ash from household chimneys and

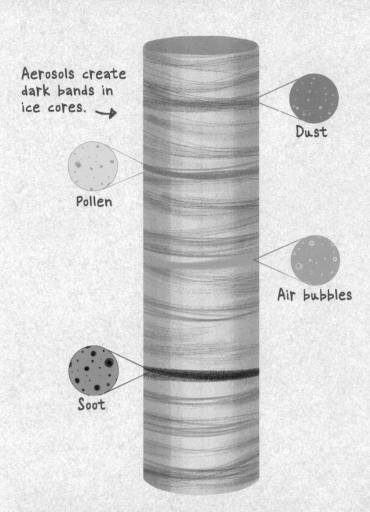

Aerosols create dark bands in ice cores. →

Dust

Pollen

Air bubbles

Soot

factory furnaces. The black streaks tell an ominous story, revealing the immense increase in ash and carbon entering the atmosphere beginning around 1750.

Another piece of the puzzle provided by ice cores is information about past temperatures. Snowflakes have a different chemical structure depending on the temperature of the air when they formed. So by looking at the ancient snow preserved in ice cores, scientists can estimate how cold it was in winters long past.

REFRIGERATING HISTORY

Critical clues that might help us solve the riddles of climate are found in the very glaciers that may soon disappear. To keep from losing this priceless data, ice cores are being refrigerated. At the National Science Foundation Ice Core Facility near Denver, Colorado, ice cores from all over the world are stored in metal containers, piled on shelves in a giant walk-in freezer. They're kept at a steady temperature of –36°F (–37.78°C). Students and scientists from all over the world go there to use the cores in their research.

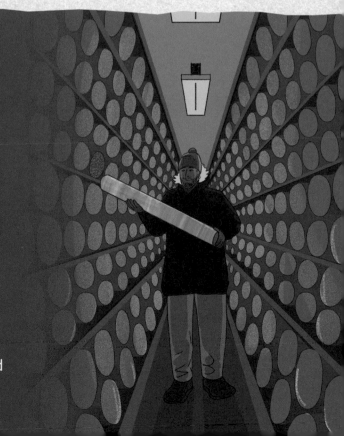

What Do the Glaciers Tell Us About CO_2?

What other secrets do glaciers hold? It turns out that glacier ice contains exactly the treasure that climatologists were seeking—actual air samples from the distant past.

As glaciers form and each snow layer is overlaid by more snow, a little bit of air is trapped in small pockets between the snowflakes. Even the hardest, oldest glacial ice isn't solid like a rock. It's filled with tiny air bubbles, like a piece of bread. This air is then buried deeper and deeper in the glacier as time passes and snow packs on top of it. Air bubbles have been buried in the oldest glaciers for millennia, and they are like time capsules—tiny breaths of air from past eons.

To extract the air samples from the core, a chunk of ice is placed under a vacuum hood, which keeps other air out. Carefully crushing the ice, chemists capture these tiny samples of gas, then analyze them to measure precisely how much carbon dioxide they hold. (CO_2 and other gases are measured in parts per million, abbreviated as *ppm*.) This analysis reveals how much CO_2 was in Earth's atmosphere in past ages and can be compared to levels of CO_2 in modern times.

The cores show that CO_2 levels in the air have gone up and down over time, as forest fires, volcanic eruptions, and natural emissions from plants and animals put carbon in the air. For example, 350,000 years ago the carbon dioxide level was 180ppm. It took about 50,000 years for it to rise to 280ppm, then over the course of another 50,000 years, it sank back down again. Carbon dioxide levels have seesawed up and down for hundreds of thousands of years, never rising above 300ppm—until the Industrial Revolution came along.

At the start of the Industrial Revolution around 1750, when cores show the dark bands of industrial pollution, they also show an abrupt rise in greenhouse gases—not only CO_2, but also N_2O and methane. After 1750, CO_2 levels in the atmosphere rose rapidly.

The cores also reveal a frightening truth: Carbon dioxide is now at its highest level in almost a million years. Today, the CO_2 in Earth's atmosphere is well above 400ppm. And it's still rising.

CONFUSING DATA

There's a weird blip on some graphs tracking temperature and CO_2 in past centuries. Information from the cores sometimes shows that at the end of an ice age, the level of CO_2 goes up *after* the temperature goes up. At first glance, this seems strange. If rising CO_2 levels cause temperature increases, shouldn't the CO_2 levels rise first? But it's more complex than that.

At the end of an ice age, the planet warms because of changes in ocean circulation and Earth's orbit, and increased solar radiation. As the planet heats up, the oceans heat up. Warmer water holds less CO_2, so this powerful greenhouse gas escapes from the oceans into the air. So on graphs of these eras, CO_2 rises after the temperature does.

But while CO_2 didn't cause the warming trends in these long-ago times, the CO_2 from the oceans increased the greenhouse effect enormously. As the temperature rose, more CO_2 was released from the oceans, making the air hotter. And remember, once warming starts, it accelerates—as darker soil and water are exposed, the air temperature rises even higher in a cycle of ever-increasing warming.

Piecing Together the Puzzle

Because the ice cores reveal both the temperatures and the CO_2 levels of past eons, we can make graphs showing temperature over time and comparing it to the level of CO_2 in the air. Such graphs show that carbon dioxide and temperature have risen and fallen together, closely linked over immense periods of time.

The ice cores show that Earth has had many periods of warming and cooling throughout its history. Climate-change doubters sometimes point to the fact that ice ages have come and gone, and glaciers have melted in the past—so what's the big deal if they're melting now?

But there's a crucial difference between historical ice ages and what's happening today. The melting of glaciers long ago was caused by various outside factors, including solar activity and changes in Earth's axis and orbit—for example, a tiny change in the tilt of the earth toward the sun changes the amount of solar heat that reaches Earth and triggers an ice age.

But in the past few decades, scientists have not observed any changes to the sun or to the earth's orbit that would account for such a rapid temperature increase. What we're experiencing now isn't just another one of Earth's natural warming periods. Something else is causing the planet's temperature to rise.

The history hidden in the glaciers plainly shows that fossil fuel emissions from factories, power plants, and motor vehicles have poured massive amounts of greenhouse gases into our air since 1750. And scientists have known for more than two hundred years about the greenhouse effect—that CO_2 and other gases cause planetary temperatures to rise.

Science tells us that glaciers are melting because of climate change related to human activity.

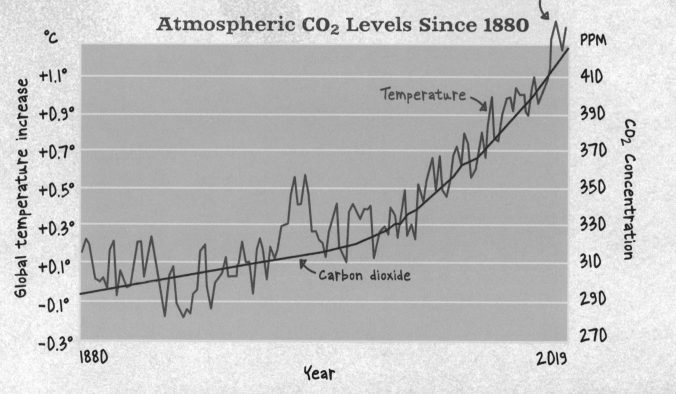

CO_2 levels have risen sharply over time along with global temperatures.

Atmospheric CO$_2$ Levels Since 1880

°C

Global temperature increase

+1.1°
+0.9°
+0.7°
+0.5°
+0.3°
+0.1°
−0.1°
−0.3°

Temperature

Carbon dioxide

PPM

410
390
370
350
330
310
290
270

CO$_2$ Concentration

1880   Year   2019

WHAT'S THE CO$_2$ LEVEL TODAY?

To find out how much CO_2 is in Earth's atmosphere right now, visit CO$_2$.Earth at **www.co2.earth**. You can find daily CO_2 levels, as well as records for past years. Try comparing this month's CO_2 levels to the same month in previous years.

Why Care About Melting Glaciers?

When you hike a glacier, it may feel like you're the only living thing for miles. So what does it really matter if glaciers melt away? After all, hardly anyone goes to these remote places, and it seems like nothing lives there. Aren't they just wastelands of empty ice and snow?

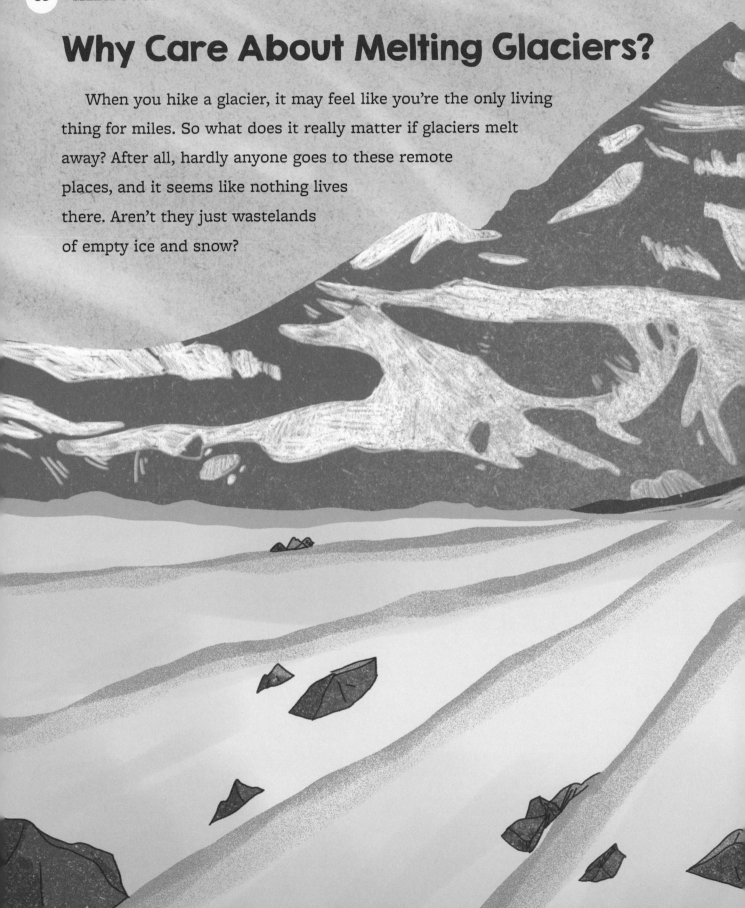

But when you walk on a glacier, you're never alone.

Over your head, under your feet, and hidden all around you are living things: plants and animals that thrive on cold.

We usually think of cold as something bad, something we want to get rid of. After all, humans are warm-blooded mammals—we like to be warm. Giving off body heat is the sign that an animal or a person is alive. "Cold as death," we say.

But on the glacier,

cold is life.

Chapter 5

Some Like It Cold

Animals and Plants of the Glaciers

If you put your hand into water flowing from a glacier on a warm summer day you'll feel the cold bite your fingers. Stick a foot in the water and your toes will turn red and ache with cold. If you dare to jump in, you'll jump out fast, tingling all over and gasping for breath. Glacier water is *cold*.

In parts of the world where there are no glaciers, waterways often dry up completely in late summer, leaving only dusty streambeds till spring comes again. In glacier country, though, the opposite happens. Instead of summer being the driest time, the hot summer sun causes an abundance of fresh water, almost as cold as the glacier it just melted from, to pour into streams and rivers. The water begins to chatter and gurgle as it flows over its stony bed.

Just in time for the salmon run.

Imagine you're walking along a glacial river in the Pacific Northwest in summertime. The water's surface ripples in a strange way. Then a flash of red catches your eye. A fish erupts from the water, curving through the air. *Splash!* Back in the water, the fish swims on, and it's not alone. The clear, icy water is filled with salmon— thousands of them, all heading upstream.

Wait. Swimming *upstream*?

The Keystone Fish

Salmon start life in freshwater streams, but they soon head for the rich food sources found in the ocean. They feed and grow in salt water, sometimes for years. (How long they stay in the ocean depends on the species). But at some point in their lives, they return home to the freshwater stream to breed, swimming upstream against the current—that is, if they manage to survive the journey.

Look! An osprey swoops low, snagging a fish with its sharp talons. And that's just the beginning of the banquet. Bald eagles hunt for fishy prey. River otters pounce on salmon in deep pools. Then the bears arrive.

Grizzlies line the riverbanks, needing to fatten up before the long winter ahead. A mother bear splashes into the water as hungry cubs watch, then comes out dripping, a wriggling salmon in her mouth. The cubs shove one another in their rush

to be first to grab a mouthful. Another bear uses four-inch (10.16 cm) claws to snag fish after fish, scooping them out of the water to lie flopping on the bank.

{Food Chain}
The transfer of energy from one living thing to another.

When eagles, bears, and otters are done feasting, their leftovers lie scattered along the shore. That's not the last stop on the **food chain**, though. Not one molecule of salmon goes to waste. Scavengers like foxes, crows, and ravens eat their fill. Worms, snails, and insects join in, until finally, all traces of the fish are gone. Even the salmon bones decompose into forest soil, adding rich nutrients like calcium and nitrogen that give life to the immense trees that tower overhead.

Salmon are a **keystone species** in the Pacific Northwest and many other places, meaning their bodies provide nutrition for a huge web of other living things.

{Keystone species}
A species that is essential to the survival of many other species in its environment.

Much of the rich tapestry of life found near glaciers in these regions depends on the salmon in one way or another. But the salmon are at risk. Salmon are **chionophiles**: animals that thrive in cold weather. In all stages of their lives, from egg to adult, salmon need cold glacial water to survive. Their ideal temperature is approximately 40–50°F (4.44–10°C). If water is a lot warmer for a prolonged length of time, it's deadly to them.

Why? Fish don't breathe water; they breathe molecules of oxygen (O_2) dissolved in water. The colder the water, the tighter its molecules are packed together, and the more oxygen stays in place. But in warmer water, molecules are farther apart, letting oxygen escape into the air and lowering the water's oxygen content. The salmon die from the lack of oxygen, drowning in the tepid water.

A food chain in the Pacific Northwest ➔

Even a one-degree rise in temperature is dangerous, triggering a host of chemical changes in the salmon's body. Warm water speeds up their metabolism, so the fish burn energy more quickly. Therefore, they have to eat more often and expend precious energy to find ever more food. Warmer water also hosts germs, bacteria, and fungi that cause disease. In 2015, a die-off of an estimated quarter of a million salmon choked lukewarm streams with the silvery bodies of dead and dying fish. Scientists determined that the fish likely died from a warm-water disease that causes gill rot.

As the climate warms, water gets hotter. And it's getting hotter earlier in the spring.

If the glaciers begin to melt too early in the season, the water will be many degrees warmer than normal by the time the salmon get there. There will also be less water—possibly too little for them to swim upstream during the late summer or fall runs.

So melting glaciers leads to fewer salmon. And the decline of salmon populations has an enormous effect on all those other organisms—eagles, grizzlies, spruces, and more—that depend on salmon. All of these organisms are part of the glacier **ecosystem**.

We tend to focus on the big, obvious members of any ecosystem: salmon, eagles, grizzlies. But they're not the only ones affected by the melting of their icy home. The glacial ecosystem has many organisms—some so small and seemingly insignificant that most people don't know they exist. . . .

THE PEOPLE OF THE SALMON

The lives of the people of the Swinomish Indian Tribal Community of the Pacific Northwest have always been closely interwoven with salmon. Traditionally, the annual salmon run has been a time of rejoicing, as people fish and celebrate along the icy rivers. The tribe has long relied on salmon as food, making it a central part of feasts and festivals marking every milestone of life from birth to death. The Swinomish were—and still are—known as "the People of the Salmon." But nowadays salmon are becoming frighteningly scarce.

"We were taught that if you take care of the land and the resources, the land will take care of you," said Kathryn Brigham, a member of the Umatilla tribe and one of the founders of the Columbia River Inter-Tribal Fish Commission. American Indian nations have fought to protect salmon migration routes from dams and power plants, using protests and lawsuits—but there's no way they can cool down streams and rivers if temperatures keep rising.

Swinomish people canoeing

"There's just no water. The glaciers are almost gone."
—LORRAINE LOOMIS, fisheries director for the Swinomish Indian Tribal Community

Ice Worms

Deep in glacier country, there have always been legends of mysterious creatures. There's the story of Bigfoot, a shaggy monster who leaves tracks in the snow. There are rumors of the Yeti, aka the Abominable Snowman, hunting prey with razor claws. There are tales of mysterious ice worms that tunnel deep in the blue ice of glaciers, avoiding sunlight and only appearing in darkness.

Actually, that last one is true.

If you walk on the glacier on a sunny morning, you won't see any worms crawling around. It's hard to imagine that any life exists in this expanse of bare, crunchy snow. But as the sun begins to sink, patches of shade appear. That's where you first spot the worms.

They're no bigger than an eyelash. Here's one, there's another—small brown bits of life squirming upward from the snow. Don't pick them up! The heat of your hand would kill the worm instantly.

During the day, the worms hide. No one knows how far down they can go—they've been found in ice caves thirty feet (9 m) below the surface. But when the sun sets, worms slither through cracks in the ice, flocking to the surface to feed.

Feed? Seems like there's nothing here but snow! The glacier may look as barren as rock, but for ice worms, it's a pasture. Microscopic specks of cold-hardy algae, fungi, and bacteria grow on snow crystals. Plenty of food for a hungry ice worm.

> **{Ecosystem}** A community of living things that interact with each other and with their environment. An ecosystem includes everything—every plant, animal, and microbe, from the most ferocious predator to the tiniest speck of algae. An ecosystem involves many food chains all gathered together into a complex web. But an ecosystem also includes the nonliving environment: sunlight, ice, rainfall, rocks, and climate. All these parts interact with each other and depend on one another.

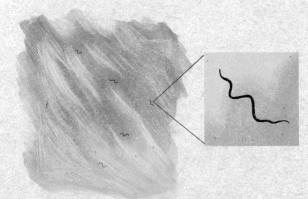

Some glaciers of the North Cascades have millions, even billions of ice worms. Biologists have counted thousands within a square foot. In some places it's impossible to walk without stepping on worms.

But ice worms don't live in plain old snow. Somehow, they sense areas that aren't underlaid by glacial ice and avoid them. For these tiny creatures, the glacier isn't just their habitat, it's the only place they can exist. And their world is getting smaller. As a glacier vanishes, its ice worms go with it.

But what does it matter, really? Only a few human beings have ever laid eyes on ice worms. What would a world without ice worms be like?

A small bird hops around the glacier in the dusk. Its plump body has chocolate-brown feathers edged with rosy pink. The bird pecks at the ice, flutters a few feet, then pecks again.

You wouldn't think a glacier would be a place for songbirds to hang out, but rosy finches are chionophiles, like salmon. They nest near glaciers, higher up in mountains than perhaps any other bird in North America. In such a harsh environment, there aren't a lot of items on their menu. The millions of ice worms provide a banquet of high-energy nutrition that lets rosy finches survive in their cold habitat. A world without ice worms would be a world with fewer birds.

Finches need ice worms to eat— that's a clear connection between two organisms in an ecosystem. But sometimes the links that hold an ecosystem together aren't that obvious.

Rosy finch

ICE WORM ANTIFREEZE

If ice worms live deep inside ice, why don't they freeze solid? One reason is that ice worms have chemicals in their bodies called antifreeze proteins (AFPs). Other animals have AFPs too, but ice worms seem to be especially good at not freezing to death.

In 2011, Canadian biochemist Mariève Desjardins began to study ice worm AFPs. To transplant an organ like a kidney from one body to another, the kidney must be kept cold without freezing, to protect delicate tissues. Even with refrigeration, a kidney can only exist outside a body for a few hours. Could ice worm AFPs help doctors keep transplant organs alive longer? No one knows yet, but it's possible.

When any living thing is driven to extinction, all of its possibilities evaporate, too, including potential benefits and life-saving cures it could offer to humans. If ice worms go extinct, the secrets of their amazing bodies will vanish with them.

For example, it might seem as if there's no connection between a tiny ice worm high on the glacier and a hungry grizzly far downstream. But let's look closer.

After an ice worm dies, its body decomposes, leaving specks of nutrients on the snow. When biologists analyzed glacial meltwater, they found it contained high amounts of these nutrients, flowing downstream to nourish plants, which are eaten by insects, which are eaten by salmon. Which are eaten by grizzly bears.

We tend to examine each part of an ecosystem one at a time, looking at a lone plant or animal as though it could survive in a vacuum all by itself. But no organism exists on its own. In an ecosystem, they all work together and depend on each other, like parts of a body. The heart can't beat without the lungs, which can't breathe without blood, which can't flow without veins. A glacier is like a living organism, made up of animals, plants, water, rock, and air. And none of it can exist without cold.

What About Us?

As glaciers shrink, all the animals depending upon them for their survival—from tiny ice worms to massive grizzlies—are threatened. If the meltdown doesn't stop, these animals will soon be pushed to the verge of extinction.

LIFE IN THE COLD

Countless species of wildlife and even plants are at risk of extinction as glaciers vanish. Here are a few other living things that rely on glaciers in ways you might not expect:

Mountain Goats: Mountain goats' thick fur is incredibly good insulation. Temperature far below freezing? No problem to a mountain goat. Problem is, they can't unzip that coat and take it off when the weather turns warm. Mountain goats get dangerously overheated if a hot spell comes along. If they can't lower their body temperature, they'll die. So the goats head for glaciers to seek relief from heat and pesky insects. Cold air from the glacier flows downslope, lowering the temperature for miles around like a giant air conditioner.

Glaciers can be a lifesaver for other thick-furred animals like musk ox, reindeer, bison, and the small rabbitlike creatures known as pikas.

And there's another species of animal we can't leave out of the picture: humans. Because the climate crisis isn't just affecting the frozen parts of the world. If the glaciers keep melting and the climate keeps warming, it will affect every ecosystem and organism on the planet—including you and me.

Seals: Where's the noisiest place in the ocean? It's at land's end where glaciers meet the sea. Chunks of ice falling into the water make loud splashes. Also, bubbles of ancient air trapped in the glacier escape into the water, popping and fizzing. All that noise is a good thing for harbor seal babies. Killer whales use sound to find their prey, so in the racket whales have a hard time locating young seals. But as glaciers shrink, their edges are retreating from the sea. If things quiet down along the coasts, that makes it easier for whales to hunt—but the whales eventually lose, too. If the whales eat too many of the young harbor seals, the seal population can't recover and the whales lose a major source of food.

Glacier Mice: If you pick up a glacier mouse, you're in for a surprise. It's not an animal; it's a plant—a type of moss. Glacier mice (that's really what botanists call them!) are round, fuzzy clumps of moss the size of a mouse. They don't have roots—they roll across the glacier's surface, blown by the wind, absorbing water from the snow. Tiny insect-like creatures called springtails live in the moss, and they in turn supply food for birds and other creatures.

Chapter 6

A World Without Glaciers

What Will Happen If We Don't Change Course

For decades, glaciologists have compared glaciers to the legendary canary in the coal mine. Coal mines are dangerous places that can suddenly fill with carbon monoxide, an odorless gas deadly to humans. It was an old tradition for coal miners to bring a caged canary down into the mine with them. Birds are more sensitive than humans to the effects of air pollutants, including carbon monoxide. If the bird's song fell silent, miners would know danger was near. When the bird died, it was time to flee. Glaciers are like those canaries—they're giving us an early warning, begging us to pay attention to what's happening to the environment around us before it's too late.

We're already seeing many of the repercussions of glacial melt, and they are only going to get worse if we don't adjust course.

Shrinking Freshwater Reservoirs

If you scooped up some glacial meltwater in your hand and tried a cold sip, you would taste the purest, most refreshing water on Earth. Glaciers are gigantic reservoirs of fresh, unpolluted water. Thousands of communities depend on glaciers for drinking water and irrigating crops. Glaciers literally provide food and drink for millions of people.

In a world without glaciers, people could get some drinkable water from melting snow, of course. But snow is mostly air, so a gallon bucket filled with snow will melt down to only a few sips of water. Glacier ice is superdense, compressed water that took many, many centuries to form. A glacier holds far more water than an equal amount of snow.

And if the glaciers melt, they won't be coming back anytime soon. Remember what the ice cores told us—some glaciers are incredibly ancient. The great ice sheets of Antarctica took more than two million years to form. Who knows how many millions of years it could take to replace them? Glaciers aren't just a big pile of snow that can easily be replaced by a couple of years' snowfall. So that supply of drinking water that people have relied on for centuries is in danger of disappearing for good.

And there's another problem that affects you, even if you don't live in a community that depends on glacial melt. Trillions of tons of ice are melting each year. Meltwater is pouring off the glaciers, flowing into the sea—and the sea is rising.

A glacial meltwater reservoir in New Zealand

Sea-Level Rise

The change is slow—almost invisible—but year by year, the oceans of the world are rising as meltwater pours in. Each year, waves wash a little farther up the sand. Beaches and wetlands shrink a little more. And the rate of sea-level rise is accelerating.

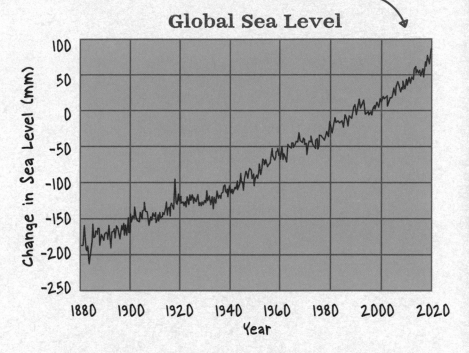

Global Sea Level

The sea level has risen 8–9 inches (21–24 centimeters) since 1880.

Many of the world's major cities are only a few feet above sea level: Lagos, Nigeria; Jakarta, Indonesia; Shanghai, China; New Orleans, USA; Amsterdam, Holland; Rio de Janeiro, Brazil. All these and many more are struggling to cope with rising waters. People who live along coasts—even in places thousands of miles away from the nearest glacier—are watching in dread as the sea slowly creeps higher. Saltwater contaminates wells of drinking water. Trees die in salt-poisoned soil. House foundations erode and collapse. And as land literally vanishes under their feet, people are forced

to leave their homes. These people, often desperate for food, clean water, and shelter, are known as **climate refugees**.

But the amount of land available for climate refugees to move to is getting smaller. Because the water is still rising.

The world's oceans have risen more than eight inches (20.3 cm) in the past hundred years and the rise is only accelerating. In the coming decades, oceans could rise by several feet. One day, cities may have to be abandoned as the millions who live there seek refuge elsewhere. And with the rising sea comes ever-increasing consequences. The melting of the glaciers affects every part of the planet—the sea, the land, the sky, and the weather they create.

South Florida in 2020, as seen from a satellite

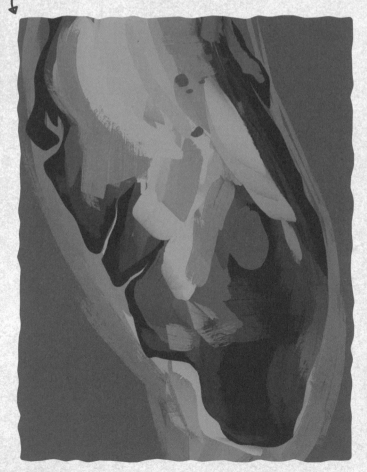

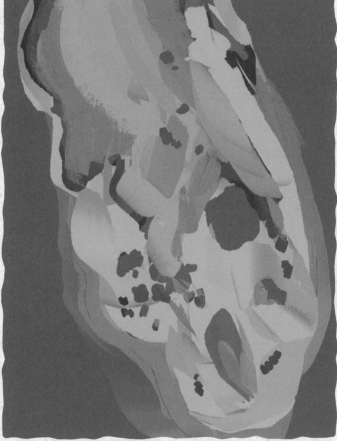

Projection of South Florida in 2100 with a 2-degree rise in global temperature

WHEN THE SEA EATS THE LAND

You wouldn't think that a tropical island with palm trees and coral reefs would be affected by faraway glaciers. But with rising sea levels, entire islands in the South Pacific have disappeared under water. Tuvalu is a tiny nation in the South Pacific made up of a string of coral islands. The highest land is fifteen feet (4.6 m) above sea level.

Already two of the country's nine islands are on the verge of disappearing under the waves. The people of Tuvalu are watching with growing despair as the Pacific Ocean devours a little bit more of their homeland every day. Most of the population is crowded onto a single island, where every inch of land is precious.

> "The sea is eating all the sand. Before, the sand used to stretch out far, and when we swam we could see the sea floor, and the coral. Now, it is cloudy all the time, and the coral is dead. Tuvalu is sinking."
>
> —LEITU FRANK, mother of five and citizen of Tuvalu

"The world wants to ignore us," says Soseala Tinilau, director of Tuvalu's Department of Environment. He's angry at the way climate change is ignored by the governments of many larger countries. "They want to keep behaving as if we don't exist, as if what's happening here isn't true. We can't let them."

Changing Currents

Climate change is not only changing the air we breathe, it's changing the water in the oceans as well. Scientists are just beginning to understand how glacial meltdown impacts the oceans of the world, which cover 70 percent of our planet.

Ocean water is in constant motion. Huge streams of water, called currents, swirl around the continents like giant conveyor belts, moving warm water from the tropics to the cold regions of the poles and back again. But floods of water pouring from the glaciers are disrupting these currents by dumping billions of tons of meltwater into the sea. Glacial melt isn't salty like seawater. It's lighter, so it floats above seawater, pushing the heavier salt-laden water down and slowing the flow of currents.

Currents affect the temperature of the landmasses they flow past. Warm currents like the Gulf Stream bring tropical warmth all the way up to Northern Europe, for example. So changes in currents will affect temperatures worldwide, heating up some areas and chilling others.

Ocean currents also shape the planet's weather systems. Warm currents heat the air above them, which changes air pressure and increases evaporation of water into the air, bringing rain and brewing storms. The world's weather is already reacting to changes in ocean currents in unpredictable ways. Storms and hurricanes are worsening around the globe. As ocean current patterns shift along with the climate, some areas get monster hurricanes, torrential downpours, and flooding. Other regions are seared by droughts, creating dangerous fire conditions.

In 2019, Australia's dry outback landscape was suffering an extreme drought. Then catastrophic fires raged across the tinder-dry land, destroying millions of acres of habitat and taking a terrible toll on Australia's unique wildlife, from koalas to kangaroos. And in a vicious cycle, smoke from the fires led to more carbon entering the atmosphere.

Storms, floods, droughts, and wildfires will only continue to intensify as the planet heats up and meltwater pours into the oceans, disrupting currents and weather patterns.

A Planet in Crisis

Glacial meltdown impacts everything on Earth. Every grizzly bear, ice worm, penguin, salmon, kangaroo, sea turtle, palm tree, and human being—all of the organisms that live on our planet are at risk as glacial ecosystems and freshwater reserves disappear, sea levels rise, and weather events become more dramatic and deadly.

And the risk is growing. Scientists studying climate change warn us of dire worst-case scenarios if greenhouse gas emissions are left unchecked: devastating droughts that make drinkable water a luxury only the wealthiest can afford. Catastrophic crop failures that lead to billions of people facing starvation. Mass extinction of millions of species of plants and animals, including a total collapse of the world's fisheries. And then there's the truly worst-case scenario: If climate change is left unchecked, Earth could ultimately become uninhabitable by humans.

What's Next?

The crisis of our changing climate can seem like a crushing weight, heavier than a glacier itself. The stakes are high. The worst-case scenarios are terrifying. Climate scientists agree that soon—*within your lifetime*—we may reach a point at which it's too late. If we keep on doing business as usual, burning cheap fossil fuels and ignoring the consequences, climate change will become irreversible. It's hard not to feel buried under the avalanche of so much bad news.

But now, get ready for some good news: Each one of us can help save the glaciers. It's not too late. If we get going now, we still have an opportunity to solve the problems that face us. We can avoid or at least limit the worst effects of climate change. The greatest danger lies in doing nothing.

It's time to take action.

REQUIEM FOR A GLACIER

Iceland is a small country, but it has some big glaciers. Hundreds of them flow from the center of the island nation toward the sea. One small glacier in the western highlands, called Okjökull (or Ok), has shrunk so much in the last few years that it has finally stopped moving. All that's left is a silent slope of black rocks and a few unmoving patches of snow and ice. Okjökull has died.

In 2019, the people of Iceland put up a tombstone for Okjökull. On it is inscribed "A Letter to the Future" written in Icelandic and English:

"Ok is the first Icelandic glacier to lose its status as a glacier.

In the next 200 years all our glaciers are expected to follow the same path.

This monument is to acknowledge that we know what is happening and what needs to be done. Only you know if we did it."

Chapter 7

A Glacier for Tomorrow

How to Take Action

Every winter, people who live near the Rhône Glacier in Switzerland spend a lot of time on their glacier. They ski it, climb it, and explore the glittering blue ice caves. And every spring for the past few years, they put their glacier to bed.

Villagers carry enormous white blankets up to the terminus and drape them over the ice. Thermal cloth reflects heat and keeps coldness inside. The blankets cover hundreds of feet of ice, but the Rhône is more than five miles (8 km) long. It's like covering up the big toe of a giant.

The Swiss hope that blankets will slow the melting of their glacier, but they know that it's not a cure, only a Band-Aid. No number of blankets can fix the cause of the problem: Our climate is in crisis.

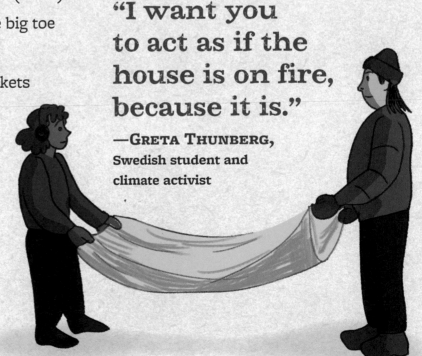

"I want you to act as if the house is on fire, because it is."

—Greta Thunberg, Swedish student and climate activist

A Planet-Sized Problem

The climate crisis seems like a problem too big for human hands to fix. How can you stop the sun from shining? How do you hold back the sea? How on earth do you save a glacier?

With any challenge, no matter how large or small, solutions begin with identifying the problem. We know that the rapid melt of the glaciers is a consequence of climate change, which is caused by our atmosphere being polluted by greenhouse gases like methane, nitrous oxide, and carbon dioxide. But who is doing all this polluting?

By far the most greenhouse gases are released by the activities of just a few dozen large corporations and power companies around the world. They mine coal, extract oil, log forests, and burn massive amounts of fossil fuels. Some of these companies are owned by the federal or state governments of powerful industrialized nations, like Russia, Saudi Arabia, and others. Some of these corporations are owned by investors—members of the public who own stock in the companies and receive profits from them. This is the case with many American corporations.

To tackle the climate crisis, we have to hold these big corporations and power companies accountable. Investors must demand that companies be environmentally responsible, and consumers like you and

me must be aware of the environmental effects of the companies we purchase from. Corporations rarely regulate themselves because regulations that protect the environment usually lead to higher production costs. That's why government regulations are essential to limit the damage caused by companies that are polluting the environment.

Sources of Greenhouse Gas Emissions in 2019

Agriculture 10%

Commercial + Residential 13%

Transportation 29%

Industry 23%

Electricity 25%

We also need to urge our governments to take action on the climate crisis. More than 120 countries have already committed to climate action. Pakistan. New Zealand. Japan. Ireland. France. Colombia. All these and many more have committed to reducing carbon emissions to zero by 2050. Each country has a different plan in place for switching to zero-carbon forms of energy, transportation, and raising food.

But tragically, many nations are ignoring the climate crisis. At the time of this writing, some of the most industrialized, biggest polluters have not taken significant action, including the United States—one of the worst offenders.

We need governments to step up and make big changes, from our local town council or mayor right up to the national level.

But how do we get this change to start?

Become a Climate Activist

It's hard to be an activist. It's scary to be the one who stands up for change. It's time-consuming to e-mail politicians and attend rallies to try to convince governments to change climate policies—and we all have lots of other things to do.

But when one person gets going, it can be the start of something big. It's like that first snowflake that falls on the mountaintop. Slowly, many snowflakes join together to create the unstoppable power of a glacier.

Here are several ways to become a climate activist.

WRITE, E-MAIL, OR CALL YOUR ELECTED REPRESENTATIVES

How do we get our government to take climate action? Well, what is a "government," anyway? It isn't some sort of machine. It's made up of individual people who, in most countries, are elected by their fellow citizens. Voters vote for the person they think can best represent them, whether in the national government or at the local village council meeting. But in order to represent you, your representatives need to know what you think. They're not mind-readers—you have to tell them.

Make it clear to your representatives that you want them to take strong action on climate change. (Generally, their contact information is easily available on the Internet.) But instead of just stating that you support environmental stuff in general, it's especially helpful to write or call with a specific goal in mind. Watch the news, read a newspaper, check reputable online sources, and talk to your family to learn about current issues. Look for a proposed issue that's about to come up for vote, and then contact your representatives to let them know how you want them to vote.

Even if you don't get to talk to officials directly, their staff will tally the number of phone calls and letters received on an issue that's about to be voted on. If no one ever bothers to write or call about climate action, what message does that send?

And don't forget to contact the elected officials in your hometown, like your town supervisor or city council representative. Or you could write a letter to the editor of your local newspaper. It can take a surprisingly small number of letters or phone calls to spur action on a local level—sometimes just one! What issues are important to you?

Like to bike? Could trails be developed in your neighborhood, so students could safely bicycle to school, reducing emissions from cars?

Love to watch birds? If there's open land in danger of being turned into a shopping mall, you might point out that a tree-filled park would help sequester carbon.

Does litter make you mad? Prod your town to be the first in the area to ban plastic bags!

What else can you think of?

VOTE!

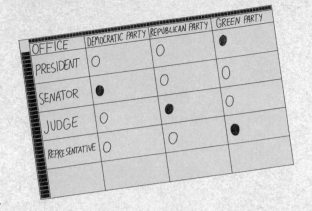

Perhaps the most important thing anyone can do for the planet is vote. Register to vote as soon as you're old enough, and support political candidates who are calling for strong climate action and who want to hold the biggest polluters—large corporations—accountable. To find out where politicians stand on climate change, read about their platforms from a nonpartisan source. The League of Women Voters has a website at **Vote411.org** that lists candidates' stances on many issues.

Even if you can't vote yet, keep insisting to candidates and your representatives that you want action on the climate crisis! And encourage every single person in your family of voting age to vote with the environment in mind.

VOLUNTEER

It can be daunting to take on climate change all by yourself. So choose one piece of the problem—a cause that's close to your heart—and find a group working to solve that problem. Many not-for-profit organizations offer hands-on volunteer opportunities. Or you could plan a fundraiser and donate the money to the organization. Here are just a few of many worthwhile not-for-profit groups working to protect the planet:

● To find out more about saving Arctic wildlife: World Wildlife Fund **www.worldwildlife.org**

● To help salmon keep swimming up those warming streams: Save Our wild Salmon (SOS) **www.wildsalmon.org**

● To learn more about planting trees and other native plants in your neighborhood: National Wildlife Federation **www.nwf.org**

SPEAK OUT

Most of all, talk about climate change. Tell everyone—family, friends, neighbors, teachers—about your ideas, fears, and hopes for our planet and for your future! Start your own environmental group in school. Use the power of social media to reach out to others. Debate solutions. Ask questions. Brainstorm. Make climate action a topic that is brought out into the open and discussed. Don't let our planet's crisis be ignored.

Reducing Your Carbon Footprint

In order to make true and lasting progress on the climate crisis, we need every government and every corporation around the world to take the crisis seriously. But while we push governments to act, we can also start making small changes to the way we live every day in order to lower our individual **carbon footprint**.

The average American is responsible for putting more than fifteen tons (13.61 metric tons) of CO_2 emissions into the atmosphere every year. Fifteen tons: That's roughly the weight of a school bus.

How can a person put pollution into the air without knowing it? If you drop litter on the ground, you can immediately see what you've done. But the environmental cost of things we do every day is often hidden.

"The climate is changing! Why aren't you?"

—GRACE OLOWOKERE, Nigerian student and climate activist

ELECTRICITY

Let's begin with something we all do every day—flipping on a light switch. At a touch, a bulb magically brightens the darkness. Where does that electricity come from?

Electricity is usually produced by fossil-fuel-powered plants, which burn coal or oil to heat water. This produces steam, which spins giant fans called turbines, creating electricity. The electrical energy travels along wires, sometimes hundreds of miles, all the way to your light bulb.

So every time you flip on a light switch, you add to the CO_2 in the air. The more electricity you use, the more fossil fuels are burned.

But there are alternative sources of energy: More and more electricity is being generated by solar panels channeling the power of the sun, or "wind farms" using giant wind-powered turbines.

What Can I Do? How many things use electricity in your house? Consider every light bulb, video game, toaster, phone charger . . . Now think about ways to cut your electric use. Realistically, it's hard to get through the day without any. You need to turn on some lights. But do you leave the TV on when no one is watching, or let lights burn in an empty room? Could you cut electricity use by 25 percent?

{Carbon footprint}
The amount of carbon dioxide released into the air because of each person's energy consumption.

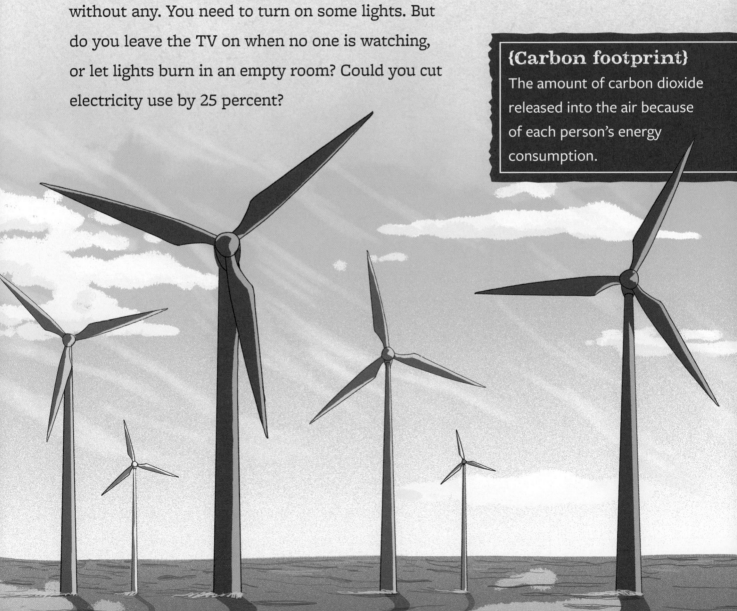

TRANSPORTATION

Go outside your house and check out the traffic whizzing by on roads and highways. Cars, trucks, vans, buses . . . they're all using fossil fuels.

Transportation is a huge cause of climate change. Gas- or diesel-burning engines give off exhaust fumes loaded with carbon dioxide. Most passenger vehicles emit about five tons (4.54 metric tons) of CO_2 per year. Heavy SUVs with poor gas mileage give off much more.

More than two million tractor trailers travel United States highways, and they give off far more CO_2 than cars do. They also puff out particles of pure carbon, called **black carbon**, which you can see in those dark clouds of smoke coming from exhaust pipes. Black carbon is a major contributing factor to climate change.

Many other greenhouse gases, such as nitrogen oxide, are also emitted by cars, buses, trucks, and planes.

So what has this got to do with you? You probably won't be driving an eighteen-wheeler down the highway anytime soon. But lots of stuff we use every day, and most of our food, is trucked hundreds or even thousands of miles, from the place where it was made to a store near you. We are the consumers of those products created by big corporations. Sneakers, cell phones, T-shirts, pencils—no matter what you buy, that thing has probably traveled many, many gas-guzzling miles to get to you.

FRESH FRUIT · EGGS · FLOWERS
PEPPERS · TOMATOES · GARLIC

Think about where your products come from. Is there anything you can buy locally, maybe at a farmers market? There's a big pollution cost to shipping an apple across the nation—buying apples from a local orchard makes for a much smaller carbon footprint. You might be thinking that if you buy one apple at the farmers market, it barely affects the pollution problem. That's true, but if a lot of us develop the habit of buying locally, the demand for apples from across the country will decrease, so there will be fewer shipments and less pollution.

Try not to waste anything, especially food. Every year, millions of tons of food are trucked across the country only to wind up being thrown into the garbage.

What if everything you bought didn't have to be new? Check out used items—clothes, toys, books—at thrift stores and garage sales. Reusing is much, much more effective at saving resources than recycling.

PLASTIC

One of the sneakiest ways pollution creeps into our world is by our use of plastic. We use plastic every day, in hundreds of ways. From your headphones to the soles of your sneakers, plastic surrounds you—but what *is* plastic, anyway?

About a hundred years ago, chemists discovered ways to rearrange the molecules in petroleum to make a new substance that's strong, long-lasting, cheap, and has a million-and-one uses. About 10 percent of the world's oil is used to manufacture plastics,

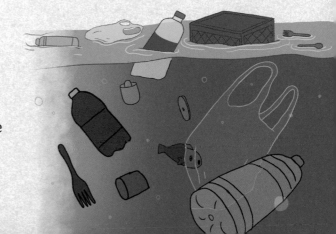

and the process gives off lots of greenhouse gases. So every plastic fork, sandwich bag, and water bottle also causes climate change and leads directly to glacial melt. Not only does the making of plastic cause pollution, but as plastic things age and begin to break down, they give off even more greenhouse gases. There's no way to use plastic without worsening climate change.

Some types of plastics can be recycled, but many can't. And even if you put it in the recycling bin, most plastic never actually *gets* recycled. The process is complex and very expensive. Most plastic ends up in the landfill—or the oceans.

What Can I Do? Identify ways you can use less plastic, especially single-use plastic—things you only use once and then throw away, like a plastic spoon. Say no to plastic bags, especially when you're just buying one or two items. Bring reusable bags to the grocery store. Use a refillable water bottle.

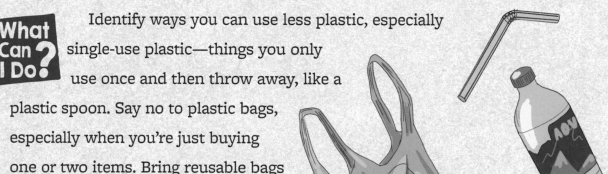

FOOD

Even what we eat makes a difference. We affect the glaciers every time we open our mouths to take a bite.

Raising any kind of food can cause the emission of greenhouse gases through the use of farm equipment and chemical fertilizers, not to mention transporting the food to your plate. But some kinds of food have a much heavier "pollution price tag" than others. Animals that we eat for food can cause immense amounts of pollution. When manure decomposes, large amounts of methane, a major greenhouse gas, are released. Also, farm animals like cows and sheep release methane from their stomachs when they burp or pass gas. More than a billion

cows live on the planet because so many people use their meat and milk for food. That adds up to a *lot* of methane, a greenhouse gas that's even more efficient at changing the climate than CO_2. Livestock contribute about 40 percent of the annual methane given off worldwide.

Raising cattle also requires a lot of open land for pastures and feedlots. Millions of acres of rainforest have been cleared to plant soybeans and other crops used to feed cattle. These cattle are mostly being raised for the fast-food industry. Millions of acres of trees that could sequester carbon and give off oxygen are destroyed when forests are cleared to make cheap burgers.

What Can I Do? Eating less meat and more plant-based food lowers your carbon footprint. Plant-based foods like nuts, beans, peanuts, and tofu all provide protein at a much lower environmental cost.

But to many people, steak is tastier than tofu! If you do eat meat, it's much better to choose meat from animals raised on small, local farms than to eat a fast-food burger.

Working Toward a Sustainable Future

Turning off a light switch, refusing a straw, and choosing a veggie burger are great ways to start—but lifestyle changes alone can't save the glaciers and the climate. If we want to fight climate change, we need planet-sized thinking. We must move toward a world where our homes, cities, cars, food—everything we do or use or eat or wear or build—is **sustainable**.

Can you switch your home's source of energy so that it doesn't come from fossil fuels but from a carbon-free source of power like solar or wind?

Could your family's next car be a hybrid or electric vehicle? Hybrids, which use both gasoline and the electrical power created by friction whenever you hit the brakes, could prevent a lot of pollution (and save money!).

When it's time to replace the fridge or the dishwasher, encourage your family to seek out "green" appliances that use less electricity. Air conditioners in particular are major sources of climate change, not just because they're electricity hogs. The chemicals used to cool the air, called HFCs (hydrofluorocarbons), are among the most powerful greenhouse gases. When old air conditioners are junked, these toxic chemicals can leak out. If your family is getting rid of an old air conditioner, encourage them to research how to dispose of it safely.

Brainstorm with the whole family—what else can you think of?

AND NOW, THINK BIGGER

Imagine you have a blank piece of paper, a lot of colored markers, and the power to design anything you want. How would you create a pollution-free, sustainable world, filled with thriving forests, clean air, and mountains covered with glaciers?

Imagine a city with no cars, filled with people walking and biking. Envision skyscrapers covered with green plants and roofed with trees. Imagine buildings heated by the sun and cooled by water or wind.

What if you could invent a kind of plastic made from grass instead of fossil fuels, that's water-soluble so that litter melts away after a rain? What if there was something you could feed cows to make them produce less methane? What if . . .

Close your eyes. What else can you imagine?

{Sustainable} Something that uses resources without using them up or causing environmental damage. A sustainable activity meets our needs but can be done for a long time by future generations.

IT ISN'T SCIENCE FICTION

The technology already exists to turn many of these ideas into reality.

For example, architects around the world are already designing net-zero buildings—homes, stores, schools, or offices that generate as much energy as they use. The first step is to lower the amount of energy the building needs: putting in skylights, insulating, using natural sun and shade. Rooftop solar panels or wind generators create energy to run lights, elevators, and plumbing. A team of researchers in Singapore is working on air-conditioning that relies on the natural cooling effect of water evaporation and doesn't use toxic chemicals.

Chemists are finding ways to make plastics from plants like potatoes, corn, or orange peels instead of fossil fuels. Some biodegradable plastics, which break down harmlessly when composted, are already on the market.

"The climate crisis is the battle of our time, and we can win."

—AL GORE, American climate activist

Solar panels harness sunlight as a source of energy, reducing or eliminating a building's need for fossil-fuel-based energy.

Fires have destroyed millions of acres of rainforest. What if we could create fireproof forests—underwater? Marine biologists in California are experimenting with farming seaweed. Underwater "forests" of fast-growing kelp, seagrass, and other types of seaweed are superefficient at sequestering carbon. One research team has discovered a nutritious red algae, another kind of seaweed, that cows can eat. Cows digest the algae so well that their stomachs produce far less methane.

Dream big! What will you invent?

GREENING THE PLANET

Green plants can change the climate of our entire planet. Just one silver maple sapling will absorb and sequester 400 pounds (181 kg) of carbon as it grows over the course of twenty-five years, keeping the carbon safely out of the atmosphere.

But, just as merely turning off a light switch won't solve the climate crisis, planting a lone tree won't, either. It takes hundreds of trees to absorb the carbon that just *one* American puts into the atmosphere each year—and there are billions of people in the world.

Earth needs **reforestation** and **afforestation** on a global scale. Dendrologists (tree scientists) estimate that it will take hundreds of billions of trees to slow climate change. One study estimated that 2.5 billion acres (one billion hectares) of additional forests would be needed—an area about the size of the United States. Planting that much forest could remove as much as 220 billion tons (200 billion metric tons) of carbon from the atmosphere. That's roughly two-thirds of the carbon humans have put into the air since the Industrial Revolution.

As well as protecting the forests we have, we need to reforest areas that have been clear-cut or burned. Then we need to afforest millions of acres, planting trees in all sorts of places: suburbs, pastures, city rooftops, deserts, mountainsides, highway medians . . . Where else?

{Reforestation}
The replanting of trees on land that used to be forested.

{Afforestation}
The planting of new forests in areas that were not formerly forests.

Why Is Change So Hard?

Why is it so hard to get people to care about climate change? Maybe it's all just too scary, too overwhelming.

It's always tempting to ignore the fact that bad stuff lies ahead. Ever have something coming up that you dreaded—a superhard test, a recital, or a big game? Instead of doing something about it that would help, like practicing, you might have just tried to forget about it—pretend it wasn't happening. It's a basic human instinct called denial. Many people don't want to think about climate change.

And there's another problem—fighting climate change doesn't come cheap.

PAYING THE PRICE

If you were shopping for a light bulb, would you buy the cheapest one, or one that costs three times as much? Seems like it makes sense to go cheap.

Do a little research, though, and it turns out the more expensive bulb is an LED (Light Emitting Diode), which uses much less electricity than regular bulbs. And electricity costs money. In the long run, you'll save by buying the more expensive bulb. And as an added bonus, you'd lower your carbon footprint.

That's the problem with climate change solutions. They cost money up front, so we look at them and say, "Nope, too expensive, can't afford that!" Redesigning a city so that it includes sustainable homes, schools, and factories could cost billions of dollars. Net-zero buildings, solar panels, and wind farms take money to build. It's expensive to buy land to plant new forests, not to mention buying seeds and tree saplings—hundreds of billions of them!

We're used to fossil-fuel energy being cheap, so cheap that we don't give the cost much thought. (When was the last time someone urged you to turn off the lights to save money?) Abandoning cheap fossil fuels will affect everyone in the world financially—including huge corporations that make billions of dollars in profits and must be held accountable to repair the environmental damage they cause.

The bottom line: Change is expensive. But spending money to fight climate change will be cheaper for all of us in the long run. Because the catastrophe of climate change and glacial meltdown is even more expensive than its solutions.

Climate change has hidden costs. What's the price tag for repairing a flood-damaged community? How much money will it take to find sources of drinking water for countries ravaged by drought? How much will it cost to feed and find homes for billions of climate refugees?

And then there's the ultimate price: a planet that is uninhabitable by humans.

The Thunberg Effect

In the summer of 2018, glaciers in Sweden were melting much faster than usual. It was the hottest summer on record—the highest recorded temperatures in more than 250 years. Wildfires broke out across the dry, overheated land.

As the summer waned, a Swedish ninth-grader named Greta Thunberg decided she had something more important to do than going to school.

Greta painted a sign with the words *Skolstrejk för Klimatet* (School Strike for Climate). She went to the Swedish Parliament building and stood outside, holding her sign and passing out leaflets about climate change. She ignored people who laughed at her. She defied those who told her to behave herself and go back to school. She refused to leave the problem to adults to solve. Alone, she stayed there day after day.

And people began to notice. First there was a post on social media. Then she was on the local news. Then all across the world, thousands were moved by her solitary protest.

Other students joined in. More and more climate strikes happened in schools all around the world. Millions of students have marched to demand that their governments do something about climate change.

Greta's voice has been heard by people of all ages who are determined to fight the climate crisis. She's spoken at climate rallies, addressed the

United Nations, and made major news headlines. In a world where it's usually adults who do the talking, Greta's words have gone viral and inspired people around the globe.

"The year 2078, I will celebrate my seventy-fifth birthday. If I have children, maybe they will spend that day with me. Maybe they will ask me about you. Maybe they will ask why you didn't do anything while there still was time to act."

—GRETA THUNBERG

Time to Begin

The Tlingit Indians of the Pacific Northwest revere glaciers, believing that the frozen giants had humanlike senses—that they could enjoy music, feel pain, or even smell food cooking. Tlingit Elders teach that glaciers could hear everything people said, and might react to their words with anger, or with sorrow. The Tlingit, like many Indigenous people, realize that humans who fail to respect nature have to face the consequences.

You've listened to the glacier, from its first whispers of dripping ice to the warning roar of the avalanche. Now it's time for you to speak. What will you say to the world about climate change?

The glaciers are listening.

Author's Note

Descriptions of glaciers in this book were inspired by Rainbow Glacier in the North Cascades in Washington State. Rainbow is one of the glaciers that Dr. Mauri Pelto's team from Nichols College in Massachusetts has monitored since 1984. It's a small glacier, covering the northeast side of Mount Baker. Its highest point is 7,200 feet (2,200 meters) and at its terminus it drains into a noisy little stream called Rainbow Creek.

In 1984, Rainbow was still touching its terminal moraine, and deep crevasses at the tip showed that it was actively moving forward. But by 1997, the terminus had retreated 738 feet (225 meters). There were few crevasses. The bulging surface became concave. By 2014, it had lost almost 40 feet (12 meters) of thickness—that's about the height of a four-story building.

Rainbow Glacier is shrinking every year. Unless we reverse the course we're on, soon Rainbow Creek will fall silent.

Rainbow Glacier in 2019

Who's Studying Glaciers and Climate Change?

Thousands of scientists with different specialties are working together to combat climate change. If you love science and you want to help save the glaciers, you might consider a career in one of these professions:

- **Biochemists** analyze the chemistry of living organisms.

- **Botanists** study plants.

- **Climatologists** try to discover how and why climate changes over time.

- **Dendrologists** specialize in the study of trees and forests.

- **Geologists** study rocks and minerals, soil, and the formation of the earth.

- **Glaciologists** specialize in glaciers, snow, and ice.

- **Meteorologists** study weather patterns and what causes them.

- **Oceanographers** focus on oceans, marine wildlife, and water chemistry.

- **Physicists** deal with how matter and energy behave, including solar radiation.

- **Wildlife biologists** study how animals interact with their ecosystems.

Additional Resources

Learn More about Glaciers:

- To see pictures of the North Cascades glaciers: North Cascade Glacier Climate Project (glaciers.nichols.edu)

- To watch glaciologist John All struggle out of the crevasse: John All, Scientist and Mountaineer (www.johnall.com/mystory)

- To listen to the songs of icebergs: LifeGate (www.lifegate.com/people/lifestyle /glaciers-melting-sounds)

- To see satellite photos of glaciers: World Glacier Monitoring Service (wgms.ch)

- To study before-and-after pictures of glacial melt over time: National Oceanic and Atmospheric Administration (www.climate.gov/news-features /understanding-climate/climate-change-glacier-mass-balance)

- To check on today's CO_2 levels in the atmosphere: CO_2.Earth (www.co2.earth)

- To track developments of glaciers worldwide: The *From A Glacier's Perspective* blog (blogs.agu.org/fromaglaciersperspective). Videos of Dr. Pelto's work on Rainbow Glacier can be found here.

- To explore the science of oceans, atmosphere, and ice: Exploratorium: Keeping an Eye On Our Changing Planet (www.exploratorium.edu/climate)

- To learn more about the science of climate change: NASA Climate Kids (climatekids.nasa.gov)

- To learn more about exciting climate change solutions: Project Drawdown: A Resource for Climate Solutions (www.drawdown.org)

- To find out about Greta Thunberg and her Friday climate strikes: Fridays For Future (www.fridaysforfuture.org)

Select Bibliography

Books

Cruikshank, Julie. *Do Glaciers Listen? Local Knowledge, Colonial Encounters, and Social Imagination*. Vancouver, Canada: UBC Press, 2005.

Gosnell, Mariana. *Ice: The Nature, the History, and the Uses of an Astonishing Substance*. New York: Alfred A. Knopf, 2005.

Hawken, Paul, ed. *Drawdown: The Most Comprehensive Plan Ever Proposed to Reverse Global Warming*. New York: Penguin Random House, 2017.

Pelto, Mauri. *Recent Climate Change Impacts on Mountain Glaciers*. Chichester, UK: Wiley Blackwell, 2017.

Streever, Bill. *Cold: Adventures in the World's Frozen Places*. New York: Little, Brown and Co., 2009.

Tyndall, John. *The Forms of Water in Clouds & Rivers, Ice & Glaciers*. New York: Appleton and Co., 1872.

Watt-Cloutier, Sheila. *The Right to Be Cold: One Woman's Fight to Protect the Arctic and Save the Planet from Climate Change*. Minneapolis: University of Minnesota Press, 2018.

White, Iain. *Environmental Planning in Context*. New York: Red Globe Press, 2015.

Articles and Websites

Chen, Jenny. "Melting Glaciers Are Wreaking Havoc on Earth's Crust." *Smithsonian Magazine*, September 1, 2016. https://www.smithsonianmag.com/science-nature/melting-glaciers-are-wreaking-havoc-earths-crust-180960226/.

Fears, Darryl. "As Salmon Vanish in the Dry Pacific Northwest, So Does Native Heritage." *Washington Post*, July 30, 2015. https://www.washingtonpost.com/national/health-science/as-salmon-vanish-in-the-dry-pacific-northwest-so-does-native-heritage/2015/07/30/2ae9f7a6-2f14-11e5-8f36-18d1d501920d_story.html.

"Home." Climate Kids. NASA, 2010. https://climatekids.nasa.gov/.

"Ice Cores and Climate Change." British Antarctic Survey. Natural Environment Research Council, February 12, 2018. https://www.bas.ac.uk/data/our-data/publication/ice-cores-and-climate-change/.

Lindsey, Rebecca. "Climate Change: Mountain Glaciers." NOAA Climate.gov, February 14, 2020. https://www.climate.gov/news-features/understanding-climate/climate-change-glacier-mass-balance.

"Overview of Greenhouse Gases." Environmental Protection Agency, 2019. https://www.epa.gov/ghgemissions/overview-greenhouse-gases.

Pelto, Mauri. North Cascade Glacier Climate Project, 2018. https://glaciers.nichols.edu/.

Rosvold, Jørgen. "Perennial Ice and Snow-Covered Land as Important Ecosystems for Birds and Mammals." *Journal of Biogeography* 43 no. 1 (January 2016): 3–12. https://onlinelibrary.wiley.xbzdoi/full/10.1111/jbi.12609.

Quote Sources

p. 17, César Portocarrero, in Casey, "In Peru's Deserts," *New York Times*, 2017.

p. 19, Deikinaak'w, in Cruikshank, *Do Glaciers Listen?*, 40.

p. 33, Mauri Pelto, personal communication, 2019.

p. 42, Svante Arrhenius, in White, *Environmental Planning in Context*, 10.

p. 45, Plato, *Critias*, trans. A. E. Taylor, 1212.

p. 66, Lorraine Loomis, in Fears, "As Salmon Vanish," *Washington Post*, 2015.

p. 77, Leitu Frank, in Roy, "Tuvalu's Sinking Islands," *The Guardian*, 2019.

p. 77, Soseala Tinilau, in Roy, "Tuvalu's Sinking Islands," *The Guardian*, 2019.

p. 83, Greta Thunberg, address at World Economic Forum: "Our House Is on Fire," 2019.

p. 90, Grace Olowokere (@GraceOlowokere), Twitter, August 2, 2019.

p. 98, Al Gore, "The Climate Crisis Is the Battle of Our Time," *New York Times*, 2019.

p. 105, Greta Thunberg, in Goodman,"You Are Stealing Our Future," *Democracy Now!*, 2018.

Graphs and Charts

p. 24: "Global Annual Mass Change of Glaciers." Graph, WGMS, 2020, World Glacier Monitoring Service.

p. 31: "Satellite Photo of San Quintin Glacier 2001–2020" In "Is San Quintin Glacier Lake the fastest expanding lake this century in South America?". From a Glacier's Perspective, December 18, 2020.

p. 42: "Atmospheric CO_2." Graph, NOAA, CO2-Earth, 2019, ProOxygen.

p. 57: "Global Temperature and Carbon Dioxide." Graph, NASA, NOAA, 2019, Climate Central.

p. 75: "Global Sea Level." Graph, NOAA, 2019, Climate.gov.

p. 76: "Florida Sea-Level Rise." https://www.miamiherald.com/opinion/op-ed/article199983139.html

p. 97: "Total US Greenhouse Gas Emissions by Economic Sector 2019." Chart, EPA, 2019, Environmental Protection Agency.

For a full bibliography, visit: workman.com/meltdownbibliography

Glossary

Ablation zone: The lower-altitude part of the glacier that loses mass due to melting, evaporation, or chunks of ice breaking off.

Accumulation zone: The upper area of the glacier where new snow accumulates and then flows down to replenish what is lost in the lower areas.

Aerosols: Particles of dust, ash, soot, pollen, or other materials picked up and blown around by the wind.

Afforestation: The planting of trees on land that was never forested or hasn't been forested for many years.

Atmosphere: The thin, life-giving layer of air that surrounds our planet. It includes the oxygen we need to survive, as well as other gases such as nitrogen, argon, hydrogen, and carbon dioxide.

Black carbon: The dark soot given off by gas and diesel engines, coal-fired power plants, and any source that burns fossil fuel. A major contributor to climate change.

Carbon dioxide (CO$_2$): The most significant greenhouse gas. CO_2 molecules contain two atoms of oxygen joined to an atom of carbon.

Carbon footprint: The amount of carbon dioxide released into the air because of each person's energy use.

Chionophile: An animal or plant that thrives in cold weather.

Climate: The overall weather pattern of an area, measured over a long time—years or even centuries.

Climate change: A significant change in the pattern of weather over a long period of time.

Climate refugee: A person who has had to leave their home because of the effects of climate change (like sea-level rise, for cxample).

Continental glaciers: Large masses of ice that cover almost all of Greenland and Antarctica.

Crevasse: A deep crack in a glacier.

Deforestation: The complete removal of trees from a wooded area, turning it into open land. This can cause soil erosion and habitat destruction.

Ecosystem: A community of living things that interact with each other and with their environment. An ecosystem involves many food chains all gathered together into a complex web, as well as the nonliving environment (sunlight, ice, rainfall, rocks, and climate, etc.).

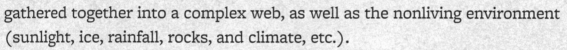

Equilibrium line: The area on a glacier at which the amount of snowfall is equal to the amount of snowmelt in a year.

Firn: Snow that has recrystallized into hard particles; an intermediate stage between snow and glacial ice.

Food chain: The transfer of energy from one living thing to another.

Fossil fuel: A substance formed in the earth from ancient animal and plant remains, which can be used as a source of energy.

Glacial flour: Dusty soil created when a glacier scrapes over rocks and grinds them into small particles.

Glacial Mass Balance (GMB): A measurement that glaciologists use to evaluate changes in a glacier's size. GMB is determined by comparing the mass gained in the accumulation zone to the mass lost in the ablation zone.

Glacier: A large mass of ice that is formed over time from compacted layers of snow and that moves slowly over the land.

Glaciologist: A scientist who studies glaciers.

Greenhouse effect: The trapping of the sun's warmth in Earth's atmosphere.

Greenhouse gas: A gas like CO_2 that increases the greenhouse effect by trapping the sun's heat close to Earth.

Ice age: A long period in Earth's history when cooling temperatures resulted in the creation of glaciers that stretched across the planet.

Iceberg: A floating chunk of ice that broke off from a glacier.

Ice cores: Long, thin samples of ice that are extracted from glaciers for study.

Keystone species: A species that is essential to an ecosystem.

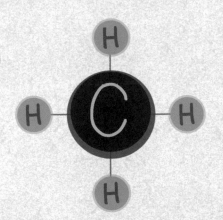

Methane: A greenhouse gas given off when something that was once alive, like a plant or an animal, decomposes.

Moraine: Soil and rocks deposited by the movement of a glacier, often forming steep ridges.

Moraine, lateral: A moraine at the glacier's side.

Moraine, terminal: A moraine at the glacier's foot, or terminus.

Mountain glaciers: Glaciers that form on mountainsides where the land is high or cold enough to have snow cover year-round. Also called alpine glaciers.

Nitrous oxide (N_2O): A greenhouse gas. Most N_2O in the atmosphere comes from nitrogen-based fertilizers used on farms and lawns.

Photosynthesis: The process by which green plants make food, using carbon dioxide, water, and light from the sun.

Radiation: The transmission of energy from one place to another in waves. Light and heat are two types of radiation.

Reforestation: The replanting of trees on land that used to be forested.

Serac: Huge blocks of ice created when a glacier flows over a steep ridge.

Sustainable: Describes something that uses resources without using them up or causing environmental damage.

Terminus: The lowest part of a glacier and the ending point of its flow.

Watermelon snow: Pink or red patches on a glacier's surface caused by algae that is cold-hardy enough to grow on ice crystals.

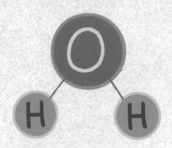

Water vapor: Water (H_2O) in gas form. As a greenhouse gas, water vapor absorbs the sun's heat and warms the earth.

Index